Buzzing Beginnings: A Beginner's Guide to Beekeeping

Learn the Art and Science of Keeping Bees in Your Backyard

Samantha Green

Table of Contents

INTRODUCTION

Welcome to "Buzzing Beginnings: A Beginner's Guide to Beekeeping: Learn the Art and Science of Keeping Bees in Your Backyard," your comprehensive manual to the fascinating world of bees and beekeeping. Whether you are motivated by a love for nature, a desire for fresh honey, or a commitment to environmental conservation, this book provides the knowledge and confidence needed to start and maintain your beekeeping journey.

Beekeeping is an ancient practice that has evolved into a blend of art and science. It offers the joy of producing your own honey and the satisfaction of contributing to the health of our planet. Bees are crucial pollinators, supporting biodiversity and the production of many of the foods we enjoy daily.

This guide is not just a theoretical exploration of beekeeping. It is a practical tool that will walk you through every step, from understanding the biology and behavior of bees to setting up your first hive and managing it throughout the seasons. Each chapter is packed with practical advice, scientific insights, and tips for troubleshooting common problems.

By the end of this book, you will not only have a solid foundation in beekeeping but also a sense of accomplishment. You will be equipped with the tools and techniques to keep your bees healthy and productive. Join us in this rewarding pursuit and discover the buzzing beginnings of your backyard beekeeping adventure.

CHAPTER I

The World of Bees

Getting To Know Bees

Bees are amazing animals that are essential to keeping our ecosystems in balance. Their most well-known function is pollination, critical in many plants' ability to reproduce and grow fruits and vegetables. Anyone interested in the environment and biodiversity, as well as beekeepers, should thoroughly understand the world of bees. This section investigates the intriguing lives of bees, looking at their behavior, biochemistry, and essential functions in the natural world.

Bees are closely related to ants and wasps and are members of the order Hymenoptera. They are divided into over 20,000 species, each suited to a particular ecological niche and environment. The three most well-known types of bees are solitary, bumble, and honeybees; they all have distinct habits and traits. The genus Apis includes honeybees, famous for producing wax and honey. They have a single queen, many worker bees, and drones living in well-organized colonies. The worker bees, who are all female, carry out different duties, such as feeding the young, tending to the hive, and foraging. The queen is the only bee that lays eggs. Drones, or male bees, primarily aim to mate with virgin queens.

Members of the genus Bombus, bumblebees are hardy, hairy, and suited to colder temperatures. Bumblebee colonies are smaller and less organized than honeybee colonies. They only store a little honey and make their nests in holes in the ground. In particular, bumblebees are excellent pollinators in areas where honeybees are less successful. They are essential for pollinating some crops, such as tomatoes and blueberries, since they can engage in "buzz pollination," which involves vibrating their flying muscles to loosen pollen.

Most species of bees are solitary, meaning they don't build colonies. Usually, each female builds her own nest, which is frequently built out of dirt, wood, or plant branches. The habits and preferences of solitary bees regarding nesting are incredibly varied. Even though they are solitary, they are vital pollinators, frequently outperforming honeybees in this regard. Two prominent examples are the leafcutter and Mason bees, renowned for their crucial pollination roles and hardworking behavior.

Bees' bodies are specially designed for their functions as pollinators. Bees have three smaller, simple eyes that sense light intensity and two substantial compound eyes that offer a broad field of view. Their touch and smell-sensitive antennae are essential for finding and navigating among flowers. The mouthparts of bees are designed for sucking and chewing, which enables them to manipulate wax and consume nectar. The head, thorax, and abdomen are the three main sections of a bee's body. Bees can fly with accuracy and agility because of the strong muscles in their thorax, which govern their legs and wings.

The reproductive and digestive systems are among the essential organs found in the bee's abdomen. The sting of female bees is a modified ovipositor. The queen bee uses her sting mainly to establish control over the worker bees, who employ it to defend the colony. Bees' sophisticated behaviors serve as a means of communication, one of the most fascinating parts of their biology. For example, honeybees use complex "waggle dances" to communicate with other members of their colony where food sources are. This dance demonstrates a fantastic degree of social structure and cooperation while transmitting information on the location and orientation of the flowers in the hive.

A bee's life cycle starts with an egg, which develops into a larva. Worker bees feed and tend to the larva, which goes through multiple molts before spinning a cocoon and maturing into a pupa. The metamorphosis that occurs during pupation makes the bee become an adult. Each stage lasts a different amount of time in other animals and is impacted by environmental variables.

One cannot stress the importance of bees in pollination. Bees move pollen grains from the male to the female portions of flowers as they forage for nectar and pollen, which helps in fertilization. The generation of seeds and fruits depends on this process, which also helps ensure plant species' genetic variety and survival. Bees are responsible for pollinating many crops humans rely on for food, making them essential to agriculture and food security.

Unfortunately, several problems with bee populations jeopardize our food supply and ecosystems. Diseases, pesticide use, climate change, and habitat loss have all led to a global drop in bee populations. For bees to survive, preserving and protecting their habitats is essential. These essential pollinators can be preserved through planting various bee-friendly gardens, cutting back on pesticide use, and encouraging sustainable agricultural methods.

Gaining insight into the world of bees enables us to recognize their significance and intricacy. Their behavior and biology reveal a complex equilibrium between cooperation and adaptability essential to our natural environment. Understanding bees will help us better support their protection and appreciate their significant influence on the environment and our lives.

Types of bees (honeybees, bumblebees, solitary bees)

As one of the most significant pollinators on the planet, bees are essential to the growth of food crops and the reproduction of several other plants. Comprehending the distinct actions, attributes, and ecological roles of the various bee species can aid us in appreciating them. This paper examines three primary bee species' unique characteristics and functions: solitary bees, bumblebees, and honeybees.

Apis mellifera, the scientific name for honeybees, is arguably the most well-known and economically significant bee species. These are social insects with thousands of worker bees, a single queen, and fewer drones living in big, well-organized colonies. The queen's primary responsibility is to lay eggs, which she does admirably quickly—she can occasionally lay more than 1,000 eggs in a single day. The foundation of the colony is the worker bees, who are all female. They carry out a variety of duties, including protecting the hive from outsiders, caring for the young, and searching for nectar and pollen. The male bees, known as drones, are only interested in mating with a virgin queen from a different hive.

Honeybees' habits and structure make them highly effective pollinators. Their "waggle dance" is a means for them to tell other bees in the colony where food sources are, and their hairy bodies are ideal for gathering pollen. This dance demonstrates a fantastic degree of social structure and collaboration while communicating information about the direction and distance of flowers. In addition, honeybees are well-known for making honey, which they do by gathering nectar from blossoms. The colony uses the honey as a food supply, particularly in the winter when foraging is impossible. Bees also make beeswax, which they utilize to build the elaborate hexagon-shaped cells of their hive.

Belonging to the genus Bombus, bumblebees are a significant bee species distinguished by their sturdy and hairy bodies. Instead of honeybees, Bumblebees build smaller colonies, usually numbering in the hundreds. These colonies are typically found in cavities in the earth, and they need to be more organized. Excellent pollinators, bumblebees are especially useful in colder locations when honeybee activity is lower. They can use a technique known as "buzz pollination," in which they shake their wings to shake pollen from flowers. Certain crops that release pollen more readily when vibrated, such as blueberries and tomatoes, are best pollinated with this technique.

Bumblebees have a different life cycle than honeybees. A queen bumblebee emerges from hibernation in the spring, locates an excellent place to build a nest, and starts laying eggs. The queen can concentrate only on laying eggs because the first brood of workers takes care of the nest's upkeep and foraging. New drones and queens are produced as autumn draws near. The old queen and the other colony members perish after mating, while the young queens hibernate.

As their name implies, solitary bees don't build colonies. Instead, every female solitary bee usually builds her nest and tends to her young. Many species of solitary bees, such as mason, leafcutter, and carpenter bees, each with their nesting habits and inclinations. For example, leafcutter bees utilize cut leaves to line their nests, while mason bees build theirs using mud found in natural cavities or hollow stems.

These bees are extremely effective pollinators, frequently surpassing honeybees in this regard despite their solitary lifestyle. Because they concentrate on a single kind of bloom on each foraging trip, solitary bees are especially good at pollinating some crops. This increases the likelihood of successful pollination. They have a significant pollination contribution and are essential to preserving the biodiversity of both cultivated and wild plants.

Every kind of bee has distinct behaviors and adaptations that help it fit into its particular ecological niche. Honeybees are now essential to agriculture and honey production because of their intricate social structure and capacity to make honey. Bumblebees are crucial for pollinating some crops and prospering in colder locations because of their solid bodies and capacity for buzz pollination. Even though they don't form social groups, solitary bees are excellent pollinators that greatly enhance the diversity and health of ecosystems.

However, all bee species are under threat from a variety of factors, including disease, pesticide use, habitat loss, and climate change. The decline in bee numbers has raised serious concerns about the potential impact on biodiversity and global food security. To secure the survival of these vital pollinators, it is imperative that we take steps to preserve and protect bee habitats. Simple actions like planting bee-friendly gardens, reducing pesticide use, and promoting sustainable farming methods can make a significant difference.

In conclusion, solitary bees, bumblebees, and honeybees all play a significant role in pollination and the maintenance of thriving ecosystems. Grasping the unique traits and actions of these bees not only allows us to appreciate their contributions to the environment but also underscores the urgency of protecting them. It is our collective responsibility to address the challenges bee populations are facing to ensure their survival and, in turn, the health of our planet.

Anatomy of a bee

A miracle of evolutionary adaptation, the bee's anatomy is admirably adapted to its pollinator's and colony inhabitants' duties. As members of the order Hymenoptera, bees possess a body structure that enables them to carry out intricate responsibilities essential to their lives and the welfare of their colonies. Knowing the anatomy of bees helps us understand their activities and critical ecological duties.

The head, thorax, and abdomen are a bee's three primary body segments. The specific organs and structures in each segment allow the bee to carry out its varied tasks.

A bee's head has several essential sensing and feeding organs. The compound eyes, composed of thousands of microscopic lenses called ommatidia, are the most distinctive characteristics. With the help of these eyes, bees can see a large area and recognize color and movement, which is crucial for finding flowers and navigating their surroundings. Bees have three simple eyes, or ocelli, on top of their heads in addition to their compound eyes. The bee uses these light-sensitive eyes to help it stay oriented while flying.

An additional important sense organ of the bee is its antennae. The antennae are very sensitive to touch and smell and are positioned on the front of the head. They are employed for environmental sensing and detecting chemical messages, or pheromones, from other bees. This sense of smell is essential for obtaining food, recognizing hive mates, and spotting danger.

With specially designed mouthparts for both sucking and chewing, bees can manipulate wax and consume nectar. The tongue, also known as the proboscis, is a long, flexible structure extending deep into flowers to retrieve nectar. The proboscis folds under the head when not in use. The mandibles, or jaws, are also utilized for transporting, fighting, cutting, and molding wax.

The muscles that regulate a bee's wings and legs are located in its thorax, the most powerful body part. The forewings and hindwings of bees are two sets of wings joined by tiny projections known as hamuli, which enable the bees to fly as a single unit. Because of its configuration, the bee can move very quickly and maneuverable, which is necessary for foraging and avoiding predators. The thorax has three pairs of legs, each with features designed for a particular function. The middle legs help in walking, the hind legs are specialized for gathering pollen, and the front legs are utilized for cleaning the antennae. Pollen baskets, or corbiculae, are concave regions encircled by stiff hairs on the hind legs of honeybees and bumblebees that are used to capture pollen pellets from flowers.

The reproductive and digestive systems, among other essential organs, are located in the divided abdomen of a bee. The sting of female bees, including queens and workers, is a modified ovipositor. In addition to being used for protection, the queen's sting is also used to sting rival queens to establish her authority. In honeybees, the glands that make beeswax are also in the abdomen. These glands secrete wax, which worker bees utilize to construct the honeycomb structure of the hive using their mandibles.

The crop, or honey stomach, is where the digestive system starts inside the belly. This is where nectar is temporarily stored while foraging. When the workers return to the hive, they share the nectar amongst them to dilute it and turn it into honey. The midgut and hindgut carry out the final stages of digestion, where waste materials are expelled, and nutrients are absorbed.

The female and male bees have different reproductive organs. The highly evolved reproductive system of queens consists of ovaries that generate eggs. A queen's ability to lay thousands of eggs over her lifespan ensures the survival and expansion of the colony. Male drones have reproductive organs designed for copulation. Their only function is to mate during a virgin queen's nuptial trip, and after that, they usually perish.

The exoskeleton of bees, composed of chitin and offers protection and structural support, is another fantastic feature of their anatomy. Its segmented, rigid outer shell permits movement and flexibility. Additionally, the exoskeleton reduces water loss, essential for preserving internal equilibrium.

The morphology of bees and their social and ecological responsibilities are closely related. Bees can carry out activities vital to the colony's life, like foraging, protecting the hive, and tending to the young because of their unique structures and organs. Their physical characteristics, especially the antennae and eyes, are also the basis for their renowned "waggle dance" and pheromone-based communication.

In summary, the anatomy of a bee demonstrates a sophisticated and highly specialized structure that enables bees to flourish in various habitats, serving as a tribute to the wonders of evolution. Every component of a bee's body, including its wings and reproductive organs, is designed to serve its functions as a social insect and pollinator. Knowing their anatomy helps us understand their behavior and ecology and highlights how vital bees are to our ecosystems and how important it is to protect them.

The life cycle of a bee

The life cycle of a bee is a marvel, a series of unique stages that are not only fascinating but also crucial to the growth and function of these vital pollinators. Understanding this life cycle provides a window into the intricate biology and behavior of bees, underscoring their role in maintaining ecological balance and supporting agricultural practices. A bee's life cycle typically unfolds in four stages: the egg, larva, pupa, and adult. Each stage, with its significant modifications and adaptations, is a testament to the wonders of nature.

The egg stage marks the start of the life cycle. A queen bee can lay Thousands of eggs over her lifetime, especially at the height of the breeding season. The worker bees carefully prepare the eggs before putting them in individual cells within the hive. The size of a rice grain, each egg is joined to the cell's bottom by a tiny quantity of glandular secretion. The eggs are fertilized or left unfertilized depending on what kind of bee will emerge. Unfertilized eggs develop into male drones, while fertilized eggs usually become female worker bees or prospective queens.

The egg hatches into a larva after roughly three days, signaling the start of the larval stage. We are living in a time of tremendous expansion and progress. The larva, which stays curled at the bottom of its cell, is a little, white, grub-like organism. Worker bees are essential during the first few days of this stage because they provide the larva with royal jelly, a nourishing material. Rich in proteins, vitamins, and other vital nutrients is royal jelly. Royal jelly is still given to future queens, but after the first few days, larvae that will become worker bees are fed bee bread, a blend of honey and pollen. The larva grows by going through multiple molts and losing its skin. It grows to its maximum size in roughly six days and needs a lot of food to support its growth.

The pupal stage, which occurs after the larval stage and is when the larva becomes an adult bee, comes next. The worker bees cap the cell with a wax cover after the larva builds a cocoon around itself. The pupa goes through metamorphosis inside the cocoon, completely rearranging its bodily structure. There are notable physiological changes associated with this period. The straightforward larval form grows into the intricate adult bee body, which has separate parts for the head, thorax, abdomen, legs, wings, and sensory organs. The bee prepares for adulthood by developing its eyes, antennae, mouthparts, and other specialized features. For worker bees, this period lasts roughly 12 days; for queens, it lasts somewhat longer; and for drones, it lasts shorter.

The adult bee's emergence marks the last phase. After completing its transformation, the young adult bee chews its way out of the capped cell and becomes a member of the hive's workers. Depending on its kind, an adult bee's duties and roles vary. The majority of bees in the hive, worker bees perform various tasks that evolve with age. Typically, young workers begin by caring for the queen and cleaning cells and larvae inside the hive. As they age, they take on more difficult responsibilities like protecting the hive, making comb, and turning nectar into honey. They eventually turn into foragers and go outside to gather water, pollen, and nectar.

Male bees called drones are only interested in mating with virgin queens. Because reproduction is their primary function, they are more significant than worker bees and do not have stingers. Drone congregation locations are where drones usually gather to wait for a queen to fly by. Typically, mating occurs during flight, which the drones only experience before they perish.

The queen bee, who lives the longest in a colony, is mainly in charge of laying eggs. During peak seasons, a queen can lay up to 2,000 eggs daily. The growth and stability of the colony depend heavily on her presence and productivity. Pheromones that queens release into the hive also aid in controlling the actions and behavior of other bees, preserving social cohesiveness and harmony.

From egg to adult, a bee's life cycle is a remarkable journey of transformation and specialization. Each phase is intricately designed to ensure the survival and function of the hive. This process not only underscores the delicate balance that bees help maintain in nature but also the complexity of insect life. As crucial pollinators, bees are indispensable to human agriculture and ecosystems, contributing to biodiversity and food production. Understanding the bee life cycle is not just about knowledge, but about our shared responsibility to protect these vital insects, ensuring their survival in the face of environmental challenges.

The role of bees in the ecosystem

Around the globe, bees are essential to preserving the harmony and health of ecosystems. They aid in reproducing numerous blooming plants, which sustain various habitats and the creatures who rely on them as primary pollinators. Bees are essential for much more than just producing honey; their work is critical to biodiversity, food security, and ecological sustainability.

Pollination is one of the most critical functions that bees perform for the ecology. The pollination process involves moving pollen from a flower's male (anther) to female (stigma) parts to facilitate fertilization and seed formation. Although the wind and other animals can pollinate many plants, bees are among the most effective pollinators because of their anatomy and behavior. As they search for nectar and pollen, their hairy bodies collect pollen grains and disperse them to other flowers. Both bees and plants gain from this symbiotic interaction; plants reproduce well, and bees acquire sustenance through nectar and pollen.

Bee pollination has a significant effect on food production. Bee pollination is responsible for around one-third of humans' food, either directly or indirectly. This covers an extensive range of nuts, seeds, fruits, and vegetables. For crops like apples, almonds, blueberries, and cucumbers to yield well and create high-quality food, bee pollination is essential. Many of these crops would experience sharp drops in output without bees, resulting in a food shortage and higher prices. Furthermore, since many of the vitamins and minerals necessary for human health are found in crops pollinated by bees, the nutritional value of food would be diminished.

In addition to helping crops, bees support the well-being of natural plant ecosystems. Bees are essential for pollinating many native plants, including wildflowers. These plants' ability to reproduce sustains the ecosystem by giving ecosystem creatures, such as insects, birds, and mammals, food and a place to live. For example, many wildlife species depend heavily on the berries and seeds bees generate from their plants. These plant-pollinator connections would be disrupted by a drop in bee populations, which would have a domino effect on the environment.

Additionally, bees contribute to the promotion of genetic diversity in populations of plants. Bees facilitate cross-pollination by mixing pollen from various plants as they travel from blossom to flower. Due to this genetic exchange, plant species are more resistant to illnesses, pests, and environmental changes. Plant populations with high genetic variety can better adjust to changing environmental conditions, which helps maintain the general stability and resilience of ecosystems.

Bees interact with other species to affect the structure and function of ecosystems and their direct role in pollination. For instance, by preserving a variety and abundance of floral supplies, healthy bee populations help sustain the numbers of other pollinators, such as butterflies and birds. As a result, different species cohabit and support one another in a more resilient and integrated ecosystem.

Bees offer a wide range of ecological functions not just in terrestrial environments. Bees pollinate plants that aid in stabilizing soil and halting erosion in riparian and wetland environments. These plants are essential for preserving the water quality because they filter contaminants and lessen sediment flow. Bees maintain the health of aquatic ecosystems and the services they offer to people and wildlife by promoting the growth of these plants.

Bee populations and the ecosystems they support are in danger due to several problems that bees confront despite their fundamental role. The use of pesticides, illnesses, climate change, and habitat loss have all significantly decreased bee populations globally. The availability of nesting places and floral resources is reduced by the loss of natural habitats brought about by urbanization, agriculture, and deforestation. It has been demonstrated that pesticides, especially neonicotinoids, damage bees by interfering with their ability to navigate, forage, and strengthen their immune systems. Plants' distribution and flowering seasons are altered by climate change, which throws off the bees' calendar and causes them to miss food sources. Bee health is also seriously hampered by parasites and diseases like the Varroa mite.

Bee population conservation necessitates coordinated action at several levels. Bees can get the necessary survival materials by preserving and rehabilitating their natural habitats. Foraging bees can be supported by planting native, bee-friendly plants in parks, gardens, and agricultural environments. The harmful effects of chemicals on bees can be lessened by reducing the usage of pesticides and implementing integrated pest management techniques. Our knowledge of bee health can be enhanced, and conservation efforts can be informed by funding research and monitoring initiatives.

In summary, bees are essential to the well-being and efficiency of ecosystems. Their function as pollinators promotes biodiversity, ecological resilience, and food production. These vital functions are significantly at risk due to the fall in bee numbers, which emphasizes the urgency of conservation measures. We may take action to save bees and guarantee the sustainability of our food systems and ecosystems for future generations by appreciating and comprehending their contributions.

CHAPTER II

Getting Started with Beekeeping

Why you should consider beekeeping

For those with an interest in agriculture, sustainability, and the natural world, beekeeping—the activity of keeping bee colonies, usually in hives—offers a unique and deeply fulfilling opportunity. The journey of beekeeping is not just about financial gain and community involvement, but also about personal growth and self-discovery. This section delves into the myriad benefits of beekeeping, illustrating why it can be a deeply rewarding and transformative pastime or career.

The beneficial effects of beekeeping on the environment are among the main arguments in favor of it. Bees are essential pollinators; their work helps many plants reproduce, including many crops necessary for human consumption. People can boost biodiversity and the health of their local ecosystems by keeping bees and helping with pollination. Bees benefit gardens and agricultural fields by significantly increasing the quantity and quality of fruits, vegetables, and flowers produced in a given area. Because of disease, pesticide use, and habitat degradation, bee populations have been dropping. Beekeepers are essential to keeping them healthy and growing. Beekeepers can contribute to preserving ecological balance and offset these unfavorable tendencies by giving their bees a secure and caring home.

Producing honey and other items related to beekeeping is a solid incentive to start beekeeping. The most well-known byproduct of beekeeping is honey, a natural sweetener with many health advantages. It has antimicrobial qualities, is high in antioxidants, and helps relieve coughs and sore throats. Harvesting honey is a common source of enjoyment for beekeepers, as it can be savored alone or relished with loved ones. Beeswax, propolis, royal jelly, and pollen are among the various essential goods that beekeeping can produce. For instance, beeswax is used in many products, such as candles, makeup, and skincare products. Natural cures include propolis, a substance that resembles resin and has therapeutic effects. Due to their well-known nutritious qualities, royal jelly and pollen are frequently used in dietary supplements and healthcare goods.

Additionally, beekeeping presents a wealth of opportunities for personal and educational development. It calls for an in-depth knowledge of biology, hive dynamics, and bee behavior. Beekeepers need to become knowledgeable about the life cycle of bees, their responsibilities within the colony, and the best ways to handle and tend to them. A deeper understanding of the interconnection of all living things and the complexity of nature is fostered by this knowledge. A contemplative and soothing hobby that offers solace from the demands of contemporary life is beekeeping. Since beekeepers must frequently check on their hives and take care of any problems that may emerge, it fosters patience, observation, and a sense of responsibility.

Financially speaking, beekeeping may be a lucrative endeavor. Beekeepers can make money by selling their extra honey and other goods and using them for personal consumption. Honey, beeswax products, and other bee-related goods can be sold in local markets, health food stores, and online. In addition, some beekeepers provide pollination services to farmers, boosting agricultural yields and generating extra income. Beekeepers have a promising market since the desire for locally produced, natural honey and beeswax products has been growing significantly. In addition, beekeeping supplies and equipment are now more widely available, making it more straightforward for novices to start with a comparatively small initial investment.

Participating in the beekeeping community is a truly enriching aspect of this endeavor. Local beekeeping societies and clubs provide a platform for beekeepers to share knowledge, experiences, and support. These associations foster a sense of camaraderie and shared purpose, as beekeepers learn from each other and build lasting friendships. Beekeepers can also extend their reach to the wider community, educating others about the importance of bees and pollination. These outreach events raise awareness and inspire others to join the cause, fostering a sense of belonging and collective action.

Moreover, beekeeping supports sustainability initiatives. People can encourage sustainable agriculture and lessen their need for chemical pesticides by engaging in ethical beekeeping. Bees naturally manage insect populations by feeding on pests and pollinating crops, minimizing the need for artificial pesticides. In addition, beekeepers can switch to organic methods and avoid dangerous chemicals that might impair bee health and the environment. This aligns with more significant initiatives to encourage sustainable living and save the environment for coming generations.

To summarize, beekeeping is a rewarding and diverse hobby with many advantages. It provides a source of natural and wholesome products, promotes environmental health through pollination, and presents chances for financial gain and personal development. Beekeeping also encourages community involvement and aids with ecological initiatives. Beekeeping, whether as a vocation or a hobby, enables people to foster biodiversity, establish positive connections with the natural world, and improve both local and global habitats. Considering these benefits, beekeeping is a valuable endeavor that can offer happiness, wisdom, and a feeling of achievement to people who take on this journey.

Common misconceptions about beekeeping

Even while beekeeping is becoming more and more fulfilling, a few widespread myths about it can discourage prospective beekeepers or cause misunderstandings about this age-old hobby. To encourage more informed decision-making and a deeper understanding of the importance of beekeeping to agriculture and the ecosystem, these myths must be addressed. This section debunks some of the most common myths around beekeeping and offers clarification on the subject.

One of the most pervasive myths regarding beekeeping is that it is challenging and requires high skill. Although beekeeping does need knowledge of bee behavior, hive management, and honey production, it is entirely doable for novices to begin with the proper assistance and materials. Numerous beekeeping societies, clubs, and online communities provide introductory classes, workshops, and mentorship programs to assist new beekeepers in getting started. Simple beekeeping tools, like frames, hive boxes, and protective clothing, are easily accessible and made to make managing bee colonies easier. Beginners can progressively develop their confidence and beekeeping skills with perseverance, a willingness to learn, and regular hive monitoring.

Another myth is that a lot of area is needed for beekeeping. In actuality, beekeeping is feasible in various environments, including rural, suburban, and urban settings. Because of their adaptability, bees may survive in multiple settings as long as they can access food and water. In particular, urban beekeeping has grown in popularity as more people realize the advantages of having pollinators in urban areas and the chance to create locally sourced honey. Beekeepers can put hives on balconies, rooftops, or community gardens if there are no local laws against it and their neighbors are understanding and cooperative.

Another myth is that beekeeping requires an excessive amount of labor and time. Even though bee colonies must be regularly monitored and cared for, beekeeping is a manageable hobby with the proper organization and planning. Most beekeeping chores, including feeding, pest control, and hive inspections, can be included in a beekeeper's daily schedule. The number of hives, the time of year, and the beekeeper's particular objectives (like producing honey or providing pollination services) can all affect how long it takes. As beekeepers gain experience, they frequently create routines that maximize efficiency and reduce disruption to their daily schedules.

Another common fear that deters potential beekeepers is the risk of bee stings. It's true that bees can sting when they feel threatened, but honeybees are generally calm and unlikely to attack unless provoked. Beekeepers can significantly reduce the risk of stings by approaching the hive with caution, handling bees gently and calmly, and using appropriate protective gear like veils and bee suits. Most beekeepers become accustomed to working with bees and learn how to avoid disturbing the colony. It's important to remember that the fear of bee stings should not overshadow the many benefits and rewards of beekeeping.

Another common misperception is that allergy sufferers should not engage in beekeeping. While many beekeepers with minor allergies successfully manage their condition by taking measures, those with severe allergies to bee stings should take caution and check with their healthcare specialists before beginning beekeeping. Wearing protective gear, carrying an EpiPen or other epinephrine auto-injector, and using antihistamines can all help reduce the chance of allergic responses. Additionally, beekeepers might select breeds recognized for their submissive areduce the chance of allergic responses. Additionally, beekeepers might select breeds recognized for their submissive and less hostile demeanor.

Finally, it's a common misconception that beekeeping is solely about harvesting honey for personal use. While honey is a valuable byproduct, beekeepers also play a crucial role in pollination activities that significantly benefit the environment and agriculture. Commercial beekeepers often lease their hives to farmers, boosting crop yields for fruits, vegetables, and nuts that rely on insect pollination. This aspect of beekeeping is not just about honey production, it's about contributing to the economy and food security, showcasing the broader impact of beekeepers beyond honey.

In conclusion, dispelling widespread misunderstandings about beekeeping is crucial to improving public awareness of this worthwhile hobby. Among the many advantages of beekeeping is its ability to promote pollination, provide honey and other hive products, and foster a connection between humans and the natural world. By busting misconceptions regarding beekeeping's intricacy, space needs, time commitment, stings, allergies, and range of activities, more people can be inspired to consider beekeeping a rewarding and beneficial hobby. In the end, beekeeping is worthwhile for individuals and communities since it supports sustainable agriculture, biodiversity protection, and environmental stewardship.

Basic requirements for beekeeping

Apiculture, or beekeeping, is a deeply fulfilling and environmentally beneficial activity that, when approached with meticulous planning and adherence to a few basic guidelines, can lead to a sense of pride and accomplishment. These regulations not only support beekeepers' prosperity and safety but also ensure the well-being and production of bee colonies. Understanding and fulfilling these fundamental requirements is not just necessary, but also a source of motivation and pride for all beekeepers, novice or expert, as they maintain healthy hives and have a truly rewarding beekeeping experience.

Apiary locations are the most critical factor in beekeeping. All year, bees need access to various plentiful sources of nectar and pollen. For continuous foraging from spring through fall, apiaries should ideally have a range of flowering plants that bloom in succession. Beekeeping can thrive in urban, suburban, or rural environments if there is a wide variety of flowers and little pesticide exposure. The apiary site should also have enough drainage to avoid water buildup around the hives, which can result in moist and unhealthy bee conditions.

Appropriate hive equipment is another vital need for beekeeping. A beehive's main parts are foundation sheets, frames, hive boxes (supers), and a hive stand. Different sizes of hive boxes are available to meet the needs of growing colonies in terms of storing honey. Bees build their combs, store their honey on frames, usually made of plastic or wood, and support the beeswax foundation sheets. By raising the hive off the ground, the hive stand offers stability and keeps moisture from penetrating within the hive. Quality hive equipment that is long-lasting, simple to construct, and appropriate for their management techniques should be purchased by beekeepers.

To protect themselves during hive inspections and other beekeeping duties, beekeepers must wear protective clothing. Beekeepers can operate comfortably and confidently among their bees while being protected from bee stings by wearing a beekeeping suit or jacket with gloves and a veil. Especially in warm weather, the beekeeper should be able to stay calm and comfortable in a suit made of lightweight, breathable cloth. To reduce the possibility of defensive behavior and stings from the bees during inspections, some beekeepers also use smoker devices, which release cold smoke.

Successful beekeeping requires regular colony management and observation. Regular hive inspections are essential for beekeepers to evaluate the well-being and output of their colonies. Beekeepers conduct inspections to look for indications of disease, pests (such as Varroa mites), and general hive health. To make sure the colony has enough food and is ready for seasonal changes, they also keep an eye on the quantity of pollen, honey, and brood—developing bees—that are stored. Effective hive management strategies, like swarm avoidance, hive splitting, and honey harvesting, contribute to sustaining colony health and output all year round.

Water supplies are necessary for beekeeping because bees need water to control the temperature in the hive, dilute honey for human consumption, and preserve general hive cleanliness. To keep bees from drowning, beekeepers should have a dependable water source next to the apiary, such as a shallow dish with clean water and floating plants or stones. Ensuring water availability lessens the chance that bees would go for less appetizing sources—like pet water bowls or swimming pools—which could result in confrontations with people.

Fundamentally, beekeepers of all experience levels must embrace the joy of learning and the sense of community in beekeeping education. A thorough understanding of bee biology, behavior, diseases, pests, and seasonal hive management strategies is not just necessary, but also a fascinating journey of discovery. Attending beekeeping classes, workshops, and seminars provided by regional agricultural extension offices, beekeeping associations, or seasoned beekeepers can be a source of engagement and connection for novice beekeepers. These courses offer practical guidance, opportunities for networking with other bee enthusiasts, and hands-on instruction, fostering a sense of community and shared learning. Expert beekeepers may enhance their beekeeping abilities and support the well-being of their colonies by staying up to date on new research, advancements in hive management, and best practices, further deepening their sense of engagement and connection.

Last but not least, the sense of responsibility and the importance of perseverance are the cornerstones of successful beekeeping. For the hive's output and the bees' welfare to be guaranteed, beekeeping demands constant attention and maintenance. Beekeepers need to be ready to put in time and effort managing their hives, getting ready for the seasons, and solving problems as they come up. The benefits of beekeeping, such as collecting honey, studying bee behavior, and aiding in pollination, are not just rewards, but also a testament to the beekeeper's commitment and dedication. The effort needed is significant, but the sense of fulfillment and the positive impact on the environment and the community make it all worthwhile.

In conclusion, establishing and maintaining thriving and fruitful bee colonies necessitates a few basic beekeeping necessities. Each prerequisite—from choosing appropriate apiary sites and purchasing equipment for the hive to guaranteeing beekeeper safety, supplying water supplies, and making educational investments—is essential to the overall success of beekeeping projects. Beekeepers may have a happy and long-lasting connection with their bees and support pollination, biodiversity protection, and the enjoyment of honey and other hive products by carefully following these guidelines.

Time commitment and seasonal considerations

Time commitment and seasonal considerations
To maintain the health and productivity of bee colonies, beekeeping, like any agricultural effort, demands a significant time commitment and close attention to seasonal concerns. Beekeepers need to be aware of the seasonal variations in the environment and the bees' natural cycles to properly manage pests, gather honey, and prepare for winter. This section examines the time commitment required for beekeeping and how seasonal factors affect hive maintenance and beekeeper activities.

The seasonal nature of beekeeping, with different chores and difficulties associated with each season, is one of its distinguishing features. The beekeeping calendar starts in the spring when colonies awaken from their winter hibernation and intensify their operations. This is a crucial time for determining winter survival rates, evaluating colony health, and, if needed, restocking food resources. Beekeepers thoroughly inspect their hives to ensure the queen is producing eggs, the brood is healthy, and the colony has enough stocks of pollen and nectar to support early spring growth. Additionally, now is the time to watch for and take care of any potential pests or illnesses, like Varroa mites, which, if ignored, can weaken colonies.

With honey production and forage height, beekeeping activities intensify as spring gives way to summer. Bees work hard to gather nectar and pollen from a wide range of blooming plants, and to store the excess honey the bees are producing, beekeepers may need to build supers or extra hive boxes. Regular hive inspections are conducted to keep an eye on population growth and hive health to ensure the colony stays robust and productive. During this period, controlling swarming behavior becomes crucial because, under the right circumstances, colonies may reproduce by dividing into two or more tiny colonies.

Winter preparations and the end of honey production define late summer and early fall. Beekeepers use supers full of capped honeycomb to gather honey, leaving adequate honey stores for the bees to survive the winter. During this time, the queen's and the colony's general health is also evaluated, as weaker colonies can require additional care or supplements to increase their food stores. Bees start to lessen their activity and prepare for winter gathering inside the hive as the temperature drops and floral supplies decrease.

For bees and beekeepers alike, winter is a time of relative inactivity, but it still calls for planning and attention. To assist bees to withstand cold temperatures and reduce winter losses, hive insulation, proper ventilation, and moisture prevention are crucial. Beekeepers can protect their hives from inclement weather using windbreaks, insulating boards, or hive coverings. To ensure the colony survives until spring, it is vital to periodically check food stocks and watch for signs of activity at the hive entrances.

While beekeeping requires varying amounts of time throughout the year, it typically requires more during the spring and summer months when things are busiest. During busy times, beekeepers usually dedicate several weekly hours to maintenance duties, hive inspections, and honey harvesting. The number of hives, the level of experience of the beekeeper, and particular management techniques are some of the variables that determine how long it will take. As they gain knowledge and experience with handling bees and efficiently maintaining hives, beginners could discover that they initially need to invest more time.

Beekeeping is not just about routine work; it also requires adaptability and quick thinking in the face of unexpected events. Severe weather, pest outbreaks, or swarming can all pose challenges that beekeepers need to be ready to address. This may involve modifying management approaches and making prompt interventions to maintain the robustness and well-being of their colonies. Seeking advice from experienced beekeepers, attending seminars or webinars, and staying updated with the latest findings and recommended beekeeping techniques are all part of this proactive approach to beekeeping.

In conclusion, to maintain the health of bee colonies and the success of beekeeping operations, beekeeping demands a significant time investment and attention to seasonal considerations. Honey production can be supported, pollinator populations can be kept healthy, and beekeepers can efficiently manage their hives by knowing the seasonal cycles of bees and the activities associated with each season. The benefits of beekeeping, such as collecting honey, studying bee behavior, and aiding in crop pollination, make the year-round time and effort required worthwhile. A successful and long-lasting relationship between beekeepers and these vital pollinators can be achieved with commitment, expertise, and a passion for bees.

CHAPTER III

Equipment and Supplies

Overview of essential beekeeping equipment

A wide range of specialized tools are needed for beekeeping to enhance the well-being and production of bee colonies, facilitate hive management, and guarantee the safety of beekeepers. All equipment used in apiculture, from protective gear and hive components to harvesting tools and additional feeding devices, is essential. An overview of crucial beekeeping tools and their significance for preserving productive beekeeping operations is given in this section.

The hive, made up of many parts intended to give bees a proper living space, is one of the most essential equipment for beekeeping. The hive contains hive boxes, sometimes called supers, which bees use as their main living space and as a place to store honey and brood (growing bees). There are many sizes of hive boxes, including medium, shallow, and deep, to meet the demands of the colony all year round. Beeswax foundation sheets are stored on frames, usually composed of plastic or wood, and fit inside hive boxes where bees construct their honeycomb. The ease with which these frames can be removed for inspection, honey extraction, and colony management improves hive maintenance while minimizing disturbance to the bees.

When conducting hive inspections and other beekeeping tasks, beekeepers must wear protective clothing to avoid bee stings and maintain their safety. A lightweight, breathable cloth is typically used for beekeeping suits and jackets, which wrap the body from the head to the ankles and come with a veil to shield the face and neck from stinging bees. Additional defense against bee stings can be obtained by wearing gloves and solid boots or shoes, particularly for novices or when handling highly protective bee colonies. Beekeepers should select sting-resistant apparel that fits well, doesn't restrict movement, and offers sufficient protection from stings.

During hive manipulations and inspections, beekeepers use a smoker to calm the bees. Filled with materials like pine needles, dried leaves, or commercial smoker fuel pellets, smokers produce cold smoke. This smoke disrupts the bees' alarm pheromones, signaling to them that there is no immediate threat. As a result, the beekeeper is less likely to encounter defensive behavior or bee stings. Smokers are invaluable for maintaining hive health, gaining access to hive interiors, and conducting routine inspections without unduly stressing the bees.

They need hive tools when beekeepers operate within the hive to adjust frames, remove honey supers, and scrape propolis or wax off hive components. Standard hive tools are the frame grip or frame lifter, which is made to lift and remove frames without harming the comb or the bees, and the hive tool or pry bar, which has a flat, thin blade for removing hive boxes and frames. These instruments are necessary to extract honey and other hive products efficiently, manage colony health, and preserve hive structure.

Beekeepers utilize supplemental feeding devices to give bees more food when nectar is scarce, bad weather or other difficult circumstances arise. Bees can receive sugar syrup or pollen substitutes safely and effectively from various feeder styles, including top feeders, entrance feeders, and frame feeders. While entry feeders are fixed to the hive entrance and let bees immediately access the meal, top feeders are positioned above the hive boxes and are loaded with sugar syrup. Frame feeders are designed to fit inside the hive and offer a regulated space where bees can eat extra food without disturbing the hive's activities or drawing in undesired pests.

Extraction equipment is necessary to extract honey from the hive frames and prepare it for sale or consumption. Honey extractors use centrifugal force to spin frames quickly to collect honey without harming the honeycomb. Extractors are available in manual or motorized versions and can hold different amounts of honey and frame sizes. Honey is extracted, bottled, and kept for later use or sale after being filtered to eliminate particles and contaminants. With the use of extracting technology, beekeepers may efficiently gather honey while maintaining the integrity of the honeycomb for future use by the bees.

In summary, necessary beekeeping equipment is critical to apiculture because it facilitates hive management, ensures beekeepers' safety, and enhances bee colonies' well-being and production. Every piece of equipment, from protective gear and hive components to smoking gadgets, hive tools, food supplies, and honey extractors, has a distinct function in ensuring the year-round success of beekeeping operations. Beekeepers can maximize productivity, reduce disruptions to bee colonies, and reap the benefits of collecting honey and other products from their hives by investing in high-quality equipment. Essential beekeeping equipment is still crucial for promoting sustainable beekeeping practices and promoting pollinator health globally as beekeeping continues to change with technology improvements and sustainable practices.

Choosing the right hive (Langstroth, Top-Bar, Warre)

Choosing the proper hive is essential for beekeepers since it affects bee colonies' management approach, honey yield, and general well-being. Langstroth, Top-Bar, and Warre are three well-liked hive designs. Each has unique benefits and concerns appropriate for specific beekeeping objectives and tastes.

In contemporary beekeeping, the "Langstroth hive" is arguably the most popular and well-known hive design. Reverend Lorenzo L. Langstroth created this hive in the middle of the 19th century. It has detachable frames that hang inside rectangular hive boxes vertically. Beeswax foundation sheets are attached to the frames so that the bees can construct their honeycomb. Because of its design, beekeepers can quickly inspect and work with frames without upsetting the colony. It makes effective hive management techniques like honey harvesting, colony expansion, and brood inspection possible. Langstroth hives' uniform frame widths facilitate equipment compatibility and make managing colonies in several hives easier. This hive's scalability, climate flexibility, and applicability for amateur and commercial beekeeping operations contribute to its appeal.

The "Top-Bar hive," on the other hand, provides a more straightforward and organic method of beekeeping. Top bar hives have long, horizontal bars or slats that form the ceiling of the hive rather than frames. Bees construct their individual, frameless comb structures from these bars by building their comb downward. Since the bees make comb based on the demands of their colony rather than following a frame, this design more closely resembles how they behave in the wild. Beekeepers interested in low-intervention and sustainable beekeeping practices prefer top bar hives. They are more straightforward to build or maintain without specialist tools and need less initial equipment expenditure. However, because the comb in Top-Bar hives is less easily detachable or standardized than in Langstroth hives, controlling and inspecting colonies in the former can be more difficult.

Understanding the bees' natural behavior and instincts is key to successful beekeeping. The Warre hive, named after its inventor, French beekeeper Emile Warre, is an alternative hive design geared toward natural beekeeping methods. Smaller boxes known as 'nadirs' are put beneath the current hive as the colony grows downward in Warre hives, which are vertical in design. By doing this, you can get bees to construct comb from top to bottom, which is how they naturally develop in tree hollows—downward. This design aims to establish environments that bolster the bees' innate instincts and well-being. Instead of using frames with foundation, they employ top bars with comb guides, which enables bees to construct naturally proportioned cells without the need for artificial foundation sheets. Bee stress throughout the winter months is decreased by this design, which increases colony warmth and insulation. Though they might need more frequent attention during inspections and hive manipulations than Langstroth hives, Warre hives are popular among beekeepers interested in sustainable methods and conserving natural bee habits.

As beekeepers, you can shape your beekeeping experience and impact the health of your colonies. Consider your experience level, preferred management style, the environment in your area, and your planned beekeeping goals while selecting the ideal hive.

Langstroth hives, with their adaptability, effectiveness, and simplicity of maintenance, are the perfect choice for those who want to maximize honey production and colony health through routine inspections and interventions. Top-Bar hives emphasize sustainability and ease of use in hive management, making them appealing to individuals seeking a more organic, low-intervention approach that fits with the bees' instincts and behavior. Warre hives offer a different kind of natural beekeeping that promotes colony health and insulation, but they need close observation and are probably best suited for beekeepers with practical hive management experience.

As beekeepers, you play a crucial role in supporting pollinator health and sustainability initiatives. The final decision of hive design should represent your principles, objectives, and dedication to the welfare of bees. Langstroth, Top-Bar, and Warre hive designs each have unique benefits and things to think about that add to the varied and changing beekeeping industry around the world. By choosing the appropriate hive and modifying management procedures accordingly, you can grow healthy, productive colonies and contribute to the health and sustainability of pollinators in your communities.

Protective gear (suits, gloves, veils)

Protective gear is necessary for beekeepers to maintain their comfort and safety when handling bees and beehives. Bees may become defensive when faced with hive inspections, honey harvesting, and pest control, all of which are part of beekeeping. To reduce the risk of bee stings and increase efficiency and confidence in beekeeping activities, it is essential to understand and use protective equipment, such as beekeeping jackets, gloves, and veils.

" Beekeeping suits" are specialty clothes that encase a beekeeper's entire body, from head to ankles, offering a physical defense against bee stings. Typically, these suits are composed of breathable, lightweight materials like cotton or polyester that provide protection and comfort while enabling airflow to keep the beekeeper cool in warmer climates. To keep bees out of the suit and gloves and to ensure a snug fit that lessens exposure to stings, the outfit has elasticated cuffs at the wrists and ankles. Thumb loops to hold sleeves in place, reinforced knees for durability, and pockets to hold personal belongings or necessary tools are standard extras seen on modern beekeeping suits. Selecting the appropriate size and fit is essential to guarantee comfort and ease of movement when working around beehives.

Another vital equipment for beekeeping safety is "gloves," which shield the hands and wrists from bee stings. Usually composed of leather or a mix of leather and canvas, beekeeping gloves provide flexibility and durability when handling hive components and dealing with bees. Leather gloves offer superior defense against stings while preserving your skill and sensitivity to detect frames and operate beekeeping equipment. Ventilated gloves, which have breathable mesh panels to improve airflow and lessen perspiration in warm weather, are preferred by confident beekeepers. It's critical that beekeepers choose gloves that fit well, offer sufficient protection, and don't interfere with their ability to do sensitive duties within the hive.

As a crucial component of beekeeping protective equipment, "veils" protect the head and neck from stings and reduce disturbances to bees while inspecting hives. Usually constructed of fine mesh or netting and fastened to a hat or hood, beekeeping veils provide breathability and visibility while keeping bees away from the beekeeper's face and head. The veil must wrap snugly around the hat or helmet to avoid openings that bees could take advantage of. Depending on the severity of their beekeeping duties, beekeepers can modify their level of protection by removing certain veils from their beekeeping suits or jackets. Hooded veils offer complete protection against bee stings while preserving comfort and mobility by adding extra coverage around the neck and shoulders.

To ensure appropriate protection against bee stings and other potential hazards, beekeepers should consider quality, comfort, and practicality while selecting protective gear. Purchasing sturdy and well-made veils, gloves, and beekeeping costumes increases safety and encourages satisfaction and confidence in beekeeping endeavors. The lifespan and efficacy of protective gear in preventing harm to beekeepers during contact with the critters are extended through appropriate care and maintenance, including routine cleaning and inspection for wear or damage.

Beyond just providing physical defense, protective gear for beekeepers is essential for fostering mental wellness and lowering stress levels, which frees them up to concentrate on the upkeep and administration of their bee colonies. Beekeepers can approach hive inspections, honey harvesting, and other beekeeping chores confidently and professionally by wearing the proper protective gear. This minimizes interruptions to bee behavior and maximizes safety for both beekeepers and bees. Protective gear continues to be a crucial component of ethical and pleasurable beekeeping operations worldwide as beekeeping advances with new tools and methods.

Tools of the trade (smoker, hive tool, bee brush)
A collection of instruments is needed for beekeeping to manage hives more efficiently, prevent disturbances for

bees, and guarantee the productivity and security of beekeepers when working with colonies. The smoker, hive tool, and bee brush are some of the most essential tools in the apiculture trade, and they each have specific uses.

The most famous and indispensable instrument beekeepers use worldwide is the " smoker." This gadget produces cold smoke using materials like pine needles, dried leaves, or commercial smoker fuel pellets. Smokers aim to soothe bees during hive inspections and manipulations by interfering with their alarm pheromones. Bees may gorge on honey in anticipation of leaving their colony when they detect smoke, which they perceive as a possible wildfire nearby. As a result of this reaction, they become less likely to sting because their focus now is on defending their food supplies rather than taking down imagined enemies. Before opening the hive, beekeepers use the smoker to waft puffs of smoke around the hive frames and into the hive entrance. When used correctly, the smoker reduces stress for bees and beekeepers during routine inspections or more intrusive hive manipulations, allowing beekeepers to operate more peacefully and efficiently.

Another essential beekeeping equipment is the " hive tool, "which is made to help handle hive parts, break apart frames, and scrape wax or propolis off hive surfaces. A flat, thin blade at one end of hive tools is usually used to pry apart hive boxes and frames without damaging the wood or comb. The hive tool may contain a hook or chisel-shaped blade on the other end for cleaning hive components, lifting frames, and freeing frames trapped with propolis. With this adaptable equipment, beekeepers may precisely and with the slightest disturbance to the bees carry out regular hive inspections, hive manipulations (like adding or removing supers), and pest management chores. Because of its durable construction and ergonomic form, the hive tool is a must-have equipment for beekeepers of all skill levels, helping them efficiently maintain their hives' health and structure throughout the beekeeping season.

The final piece of equipment needed for beekeeping is a "bee brush," which gives beekeepers a soft way to handle bees and brush them off frames or hive components when doing inspections. With bee brushes, which have long handles and soft bristles connected, beekeepers can carefully and methodically sweep bees without endangering them or putting them under unnecessary stress. When bees need to be removed from honey-filled frames for harvesting or inspection, bee brushes come in handy as they help keep the bees calm and undisturbed throughout these procedures. The bee brush's gentle bristles quickly remove bees without harming or crushing them, enabling beekeepers to work without jeopardizing the colony's general well-being. When used correctly, the bee brush minimizes disturbances to hive operations and fosters peace between beekeepers and their apiaries, all contributing to a satisfying beekeeping experience.

In conclusion, beekeepers' critical equipment includes the smoker, hive tool, and bee brush, critical to hive management, beekeeper safety, and the well-being and production of bee colonies. Beekeeper efficiency and confidence increase when a smoker uses cool smoke to calm bees during hive manipulations and inspections strategically. Beekeepers may easily do crucial activities like frame manipulation, hive maintenance, and pest management using the hive tool's precision and versatility, all while causing the least disruption to the hive's structure. Bees can be handled gently and effectively using a bee brush, which helps beekeepers work with hive components to keep bees calm and unagitated during inspections. Taken as a whole, these instruments represent the complex balancing act that apiculture practitioners must do when it comes to bee safety, pragmatism, and respect. Beekeepers may develop robust and fruitful bee colonies and build lasting and fulfilling relationships with these essential pollinators by becoming proficient with this indispensable equipment.

Recommended suppliers and costs

Making the right beekeeping equipment and bee supplier selections is essential to a profitable and pleasurable endeavor. Beekeepers can find many providers providing everything from hives and protective equipment to bees themselves. Dadant & Sons, Inc. is one of the most respected and well-known suppliers in the US. Since its founding in 1863, Dadant & Sons has built a solid reputation for offering top-notch tools, woodenware, safety gear, and extraction equipment for beekeepers. Beekeepers can easily browse and buy anything they need for their apiary operations from their comprehensive catalog and online store.

Mann Lake Ltd. is another well-known supplier; it offers a wide range of equipment and supplies for beekeeping. Mann Lake provides novice and expert beekeepers with various hive parts, safety equipment, bee feeds, and harvesting instruments. They are an excellent resource for beekeepers seeking advice and knowledge because they offer workshops, educational materials, and support services.

Reputable vendors offering package bees, nucs (nucleus colonies), and queen bees for sale include Kelley Beekeeping and Brushy Mountain Bee Farm. A package of bees usually consists of a mated queen and several pounds of worker bees, ready to start a new colony in a beekeeper's hive. Nucs provide beekeepers with a more advanced starting point because they are small, established colonies with a laying queen, brood frames, bees, and food storage. To ensure robust, resilient colonies that can flourish in various environmental circumstances, it is imperative to select a supplier who prioritizes the health and quality of bees.

Cost factors change based on the provider, the area, and the particular goods or services needed. The cost of basic beekeeping supplies, such as smokers, brushes, and hive tools, can range from $10 to $50 per item, depending on brand and quality. Beekeeping suits, gloves, and veils are examples of protective gear. These items vary from $50 to $200 or more, depending on the material, style, and degree of protection provided. The cost of hive components, like frames, foundation sheets, and hive boxes, varies according to size, quantity, and material (wood or plastic). Starter kits can cost as much as $300 or more.

Bees themselves vary significantly in price as well. Package bees typically cost between $100 and $150 for each package. However, nucs are more expensive, ranging from $150 to $250, depending on the colony's size, strength, and genetic makeup. Depending on the breed and quality, queen bees can cost anywhere from $20 to $40 per queen when purchased separately. Beekeepers should set aside money in their budgets for hive upkeep, additional feeding, pest control, and equipment for extracting honey, in addition to the initial expense of equipment and bees.

Investigating and comparing providers, reading reviews, and asking seasoned beekeepers or local beekeeping associations for recommendations before buying beekeeping materials and bees is wise. Selecting reliable providers guarantees high-quality supplies and assistance, both necessary for beekeeping activities to succeed and last. Beekeepers may create and maintain healthy colonies, get the benefits of producing honey, and support pollinator conservation and well-being in their local communities by investing in dependable equipment and healthy bees from reputable providers.

CHAPTER IV

Setting Up Your Hive

Selecting a location for your hive

The health and production of your bee colony and your general beekeeping experience can be significantly impacted by choosing the best location for your hive. When selecting a location, several criteria should be carefully considered to ensure the bees' safety and ease of managing the hive.

First off, the location of hives is greatly influenced by "sunlight and shade." Bees do best in areas with constant daylight, especially in colder climates, where sunshine helps control hive activity and temperature. Hives should ideally be positioned where they receive morning sun, enabling bees to warm up early and begin their foraging tasks. To avoid the hive and bees overheating, it's crucial in hotter climates to offer shade during the warmest portion of the day.

" Closeness to water sources is an additional crucial factor. Bees need access to clean water to stay hydrated and control the humidity and temperature in their hives. Make sure there are water sources nearby, like ponds, streams, or even a tiny birdbath with rocks and fresh water so they can safely perch. By keeping water close to the hive, bees can avoid making the sometimes stressful and energy-draining long trips for water.

" Forage availability" is essential to your bee colony's well-being and output. Pick a place with plenty of floral resources, such as various plants that provide pollen and nectar at different times of the year. While vast crops and wildflowers are advantages in rural settings, gardens, parks, and natural places can provide various foraging options in urban and suburban situations. Enhancing the variety and availability of feed can help bee colonies become healthier and more productive. You can research the local flora and add bee-friendly plants to your hive.

The stability and comfort of the hive depend on " protection from wind and bad weather." To reduce exposure to severe winds, place hives in protected areas, such as beside a fence, hedgerow, or structure. A hive should not be placed in a low-lying region that is likely to flood or in an area that has harsh weather often (such as frequent storms or strong winds), as these conditions might upset the hive and stress the bees.

When choosing a place for a hive, " accessibility for beekeepers" is another practical consideration. Select a location that makes it simple to maintain, inspect, and harvest honey from your hives. Ensure you have enough room to move around the hive without inconveniencing the bees or their flight pathways. If you plan to raise the hive off the ground with hive stands or pallets, be sure the hive is level and stable to avoid tilting or shifting.

Finally, consider " neighbor issues and municipal regulations" when erecting your hive. Check your local legislation, zoning laws, and homeowner association standards regarding beekeeping and hive placement. Talk to your neighbors about your ambitions to start a beekeeping business. Address any worries they may have and ensure your hive's location respects their property and daily activities.

In conclusion, carefully considering sunshine, water sources, fodder availability, wind protection, accessibility, and community issues is necessary when choosing the proper location for your beehive. Beekeepers can provide an environment that promotes robust and healthy bee colonies, optimizes foraging opportunities, and reduces the likelihood of disturbing bees or nearby residents by selecting a site well-suited for their needs. A successful and fulfilling beekeeping endeavor begins with a thoughtful hive location, which enables beekeepers to take advantage of honey production, pollination services, and a closer relationship with the fascinating world of bees.

Legal considerations and permits

To ensure local regulations are followed and to encourage appropriate apiary management, legal concerns and permissions are essential to beekeeping. Before starting a hive, beekeepers must know about local laws, rules, and permission requirements. The country, state/province, and even local municipality have different legislation, which vary greatly and consider factors such as public safety issues, hive density, and beekeeping techniques.

Beekeeping regulations sometimes control where to put hives, how far they can be from property lines or public areas, how many hives are allowed on a property, and whether or not neighbors or local authorities must be notified. Before beginning a beekeeping enterprise, certain jurisdictions may require beekeepers to get specific permits or licenses, especially in urban or heavily populated regions where beekeeping activities may affect neighbors or public safety. To guarantee continued adherence to local laws, these permits could be subject to a charge and need to be renewed regularly.

Homeowner association (HOA) policies and zoning laws may also impact beekeeping operations. Land use is regulated by zoning laws, which may also indicate whether locations are acceptable or inappropriate for beekeeping. Beekeepers must check zoning maps and speak with local planning offices to find out if beekeeping is allowed where they want to go and whether any additional requirements or limitations apply.

Homeowner groups frequently impose restrictions on the rearing of animals or conducting agricultural operations, such as beekeeping, on their lands. Specific homeowners associations (HOAs) have the authority to limit the location of hives, mandate that beekeepers notify other neighbors of their operations, or set extra standards for hive upkeep and aesthetics. Before starting a beehive, beekeepers should research the HOA's rules and covenants and ask the board for approval or clarification to avoid potential conflicts or infractions.

Regulations about beekeeping prioritize public safety, especially in light of bee stings and allergic reactions. Specific legal regimes might mandate that beekeepers construct sufficient flyway barriers or keep their hives safely away from public spaces like playgrounds, walkways, and swimming pools. In addition to ensuring that beekeeping operations do not annoy or endanger the safety of neighboring residents or tourists, this helps reduce the risk of bee stings for the general public.

Regulations about beekeeping also take the environment into account, especially when it comes to pesticide use and bee health. Certain insecticides that could kill bees or taint goods from hives are subject to recommendations or prohibitions in several jurisdictions. To preserve bee health and reduce its adverse effects on the environment, beekeepers are urged to employ integrated pest management (IPM) techniques and, whenever feasible, look for alternatives to chemical treatments.

Adherence to legal requirements and securing required permissions indicates a dedication to conscientious beekeeping methods and cultivates favorable associations with nearby residents and municipal officials. Beekeepers must be in constant contact with local authorities, HOAs, and neighbors to resolve any issues and show their commitment to ethical beekeeping. Beekeepers support efforts to safeguard pollinators and their habitats and help ensure that beekeeping remains a crucial agricultural practice by adhering to legal regulations.

In conclusion, managing permit requirements and legal issues is crucial to starting and running a beekeeping business. Beekeepers may guarantee the sustainability and legality of their apicultural operations, advance the health and safety of bees, and positively impact their communities by being aware of and abiding by local legislation. In addition to shielding beekeepers from fines or penalties, adhering to the law creates a favorable atmosphere for beekeeping to flourish, which benefits beekeepers and the larger ecosystem that depends on the pollination services that bees perform.

Preparing the site

Setting up a beehive and making it sustainable is essential to beekeeping success. This procedure involves several important factors to protect the bee colony's health and productivity and the beekeeper's safety and convenience.

First and foremost, it is crucial to " level and clear the ground" where the hive will be located. In particular, whether utilizing hive stands or pallets, the location needs to be level and sturdy to avoid the hive tipping over or moving. This reduces the possibility of harm to hive elements and disruption to the bees by guaranteeing the hive stays safe and stable.

" Accurate drainage" is yet another crucial component in site preparation. Make sure there isn't any standing water near the hive because it can harm bee health and cause infections in the hive. To enhance drainage and keep water from accumulating around the hive, consider raising the hive a little or covering it with gravel or paving stones.

Essential factors to consider while placing a hive include " orientation and solar exposure." To ensure the hive receives early morning sunshine, face the hive entrance from southeast to south. Because of their direction, bees can get active and warm up earlier in the day, promoting earlier foraging and increased output. Sufficient sunshine exposure during the day also aids in controlling hive activity and temperature, which is crucial for colony well-being and honey yield.

" Wind protection" is essential to reduce exposure to high winds, which can stress bees and disturb hive temperatures. Pick a spot protected from the principal winds, such as beside a structure, hedge, or fence. This produces a more stable microclimate surrounding the hive, acting as a natural barrier to lessen wind exposure.

" Distance from human activity" is another factor to consider when preparing the site. Even though bees are generally harmless if left alone, it's crucial to locate hives away from busy streets and places where humans frequently congregate. By doing this, the possibility of unintentional disruptions or encounters that can result in bee stings is decreased, and beekeepers and their neighbors can live in harmony.

" Forage availability" should have a role in the choice of location. Pick a place with plenty of floral resources, such as many plants that produce pollen and nectar. While agricultural products and wildflowers benefit rural settings, gardens, parks, and natural places can provide various foraging options in urban and suburban regions. Bee health, productivity, and honey output are all boosted by increasing the food available surrounding the hive, which benefits the beekeeping business.

Finally, "consider adjacent land usage and potential hazards." Keep an eye out for nearby properties and activities that could affect beekeeping, such as using pesticides, raising livestock, or conducting industrial operations. Share your intentions to keep bees with your neighbors and take safety measures to reduce any hazards to the health and welfare of the hives.

In summary, careful consideration of the following factors must be made while establishing a location for a beehive: ground clearance, leveling, drainage, sun exposure, wind protection, distance from human activity, availability of feed, and nearby land usage. Beekeepers may create the ideal habitat for healthy and productive bee colonies by designing and setting up the hive site. By doing this, beekeeping operations become more sustainable, pollination services are improved, and pollinators and their habitats are preserved. A successful beekeeping experience is facilitated by thoughtful site preparation, which gives beekeepers access to honey production, pollination assistance, and a closer understanding of the fascinating world of bees.

Installing the hive and initial setup

A critical step in beekeeping is installing a hive and starting the initial setup, which signifies the start of a colony's journey and the beekeeper's responsibility in overseeing their new apiary. This process entails several crucial steps to guarantee that the hive is appropriately built and prepared to host a robust and fruitful bee colony.

The site of the hive must be chosen carefully before installation. Several requirements, including sufficient sunlight, appropriate drainage, wind protection, and accessibility to water supplies, should be met by the location of choice. To maximize exposure to morning sunlight, which helps bees warm up early and promotes their foraging activities, the hive entrance should preferably face southeast to south. Placement of the hive away from busy roads and human activities helps to decrease interruptions and lowers the possibility of unintentional contact with bees.

Beekeepers usually assemble parts of the hive, like frames, foundation sheets, and hive boxes, before installation. This entails checking each part's quality, ensuring the frames are equipped with foundation sheets of beeswax or starter strips for developing combs and setting up any extra tools like bottom boards or hive stands. Careful assembly and preparation guarantee a sturdy hive to hold bees upon installation.

A cautious handling and acclimation process is required when moving bees into the hive, depending on whether they came from package bees, nucleus colonies, or swarms. Beekeepers usually follow specific protocols to acclimate package bees or nucleus colonies to their new hive environment. Encouraging colony establishment and comb development may involve carefully tapping bees into the hive, freeing the queen, and ensuring bees have access to food supplies like sugar syrup or pollen patties.

Giving colonies access to water and extra food during the first setup phase is essential for colony growth, particularly for recently installed packages or nucleus colonies. Beekeepers might feed their hives sugar syrup to encourage the construction of combs and ensure the bees have enough food reserves while they adjust to their new hive. Furthermore, bees can properly hydrate and control the humidity and temperature in their hives by having a nearby water source—like a shallow dish filled with rocks.

Beekeepers should carefully monitor hive activity in the first several weeks following hive installation. Beekeepers may monitor the development of the comb, determine if the queen is laying eggs, and evaluate the colony's health by routine examinations. The long-term viability of the hive depends on modifications like adding more hive boxes (supers) as the colony grows, keeping an eye out for symptoms of pests or illnesses, and modifying feeding schedules by colony requirements.

Beekeepers keep thorough records and paperwork during the installation and initial setup stages. This includes taking notes on observations made in the hive, scheduling feedings, recording the introduction of queens, and monitoring the growth and development of the colony over time. Accurate record-keeping aids in future management decisions fosters continuous learning and beekeeping practice improvement, and aids beekeepers in keeping an eye on the health and production of their hives.

Finally, talking to other beekeepers and getting guidance from seasoned professionals can offer insightful ideas and helpful support during the installation and initial setup. Beekeeping knowledge is increased, networking possibilities are created, and cooperation in improving bee health and sustainable beekeeping practices is encouraged by joining local beekeeping associations, attending workshops or courses, and establishing connections with beekeeping mentors.

In summary, the foundational phase in beekeeping involves establishing a hive and carrying out the first setup, which calls for meticulous planning, preparation, and attention to detail. Beekeepers lay the groundwork for a fruitful and fulfilling beekeeping journey by selecting the best location for their hive, putting together the necessary parts, moving the bees with skill, giving them some food and water at first, keeping an eye on the hive's conditions, keeping records, and interacting with other beekeepers. In addition to promoting honey production and pollination services, establishing a robust and fruitful bee colony also helps preserve pollinators and their vital role in ecosystems worldwide.

Introducing bees to the hive

To successfully create a healthy colony, introducing bees to the hive is an essential step in beekeeping that must be handled carefully. Whether beekeepers purchase their swarms, packages, or nucleus colonies (nucs), the introduction technique reduces stress and encourages the bees' assimilation into their new hive.

Beekeepers deliberately introduce " package bees"; these bees usually arrive in a screened box with thousands of worker bees and a queen in a separate cage. They first ensure the hive is ready for comb construction by providing the frames equipped with starting strips or beeswax foundations. Beekeepers softly mist the bees with sugar syrup before opening the box to quiet them and lessen their flying activity. After the queen cage is carefully removed from the packaging, beekeepers ensure that the queen is still alive and well. Place the queen cage between the frames to guarantee that worker bees get used to the queen's pheromones. The bees can then gradually enter the hive by gently shaking or tapping the package. Before completely opening the hive, give the bees a day to adjust to their new surroundings by covering the hive with an empty hive box or cover.

A slightly different introduction procedure is needed for " nucleus colonies (nucs)," which comprise many frames with bees, brood (developing bees), honey, and a mated queen. With extreme caution, beekeepers move the nuc to the hive's specified place, ensuring there is as little disruption as possible. The nuc frames are examined upon arrival to verify the queen's existence and the brood's health. After the worker bees have been accustomed to the queen's pheromones, beekeepers who have the queen in a separate cage may let her straight into the hive. As an alternative, the frames are cautiously placed into the hive, ensuring they are in the right spot. Giving sugar syrup or pollen patties helps colony establishment and encourages bees to start foraging and comb building, just like package bees.

" Introducing a swarm" into a hive entails catching and moving a group of bees that have abandoned their original colony to start a new one. Beekeepers use shaking branches or guiding the swarm with a bee brush to encourage the bees into a hive box gently. Frames with drawn comb or foundation are added to the hive box once the bees are inside to give it structure and entice them to stay. If she is in the swarm, the queen will usually follow the rest of the colony into the hive box. For several days, beekeepers keep an eye on the hive entrance to ensure the bees settle in and become accustomed to their new surroundings.

It takes time and careful attention to introduce bees to the hive, regardless of the approach. In the first few days after introduction, beekeepers should refrain from overly disrupting the colony to give the bees time to adjust and start constructing combs, raising young, and foraging. The colony's acceptance of the new hive and queen is shown by activity at the hive entrances, such as guard bee behavior and pollen gathering.

In summary, introducing bees to the hive is a crucial step in beekeeping that prepares the ground for the start and expansion of colonies. Beekeepers lower the chance of queen failure or absconding by using appropriate methods when introducing package bees, nucleus colonies, or swarms. To ensure that bees successfully settle into their new hive environment and develop a healthy and productive colony that supports the production of honey, pollination services, and the general sustainability of beekeeping activities, each introduction method entails a set of stages.

CHAPTER V

Bee Biology and Behavior

Understanding bee society (queen, workers, drones)

Studying the complex social structure and roles within a hive—where each individual is essential to the life and prosperity of the colony—is necessary to comprehend bee culture. The " queen bee," the mother and colony leader, is central to every hive. Her primary function is reproduction; she lays eggs that become worker bees, new queens, or drones, which are male bees. The queen bee, physically more significant than other bees, controls the colony's behavior and harmony through her pheromones, which affect activities like feeding, tending to the brood, and protecting the hive.

The most numerous and varied group of bees in the colony, the " worker bees," surround the queen. All worker bees are female and throughout their existence, they carry out a variety of responsibilities. When young, they clean cells, tend to the queen, and nurse larvae. As they age, they build hives, gather pollen and nectar, secure the hive entrance, and convert pollen and honey into food. Age and the needs of the colony dictate the functions that workers play, and each bee moves fluidly between duties to maintain the overall health of the colony.

Unlike the workers, the males in the hive are represented by " drones." Mating with a virgin queen during her nuptial trip is their primary goal. Larger than workers, with solid bodies and wide eyes, drones do not participate in hive maintenance or foraging operations. Seasonally, their numbers vary, reaching a peak during mating flights in the spring and early summer. Drones usually perish or are driven from the hive after mating since supplies grow scarcer as autumn approaches.

One example of the astonishing efficiency and organization of social insects is the division of labor inside a colony of bees. Eusociality is the structure that uses specialization and coordinated efforts to secure the colony's survival. The specialized functions of workers and drones combined with the queen's supremacy in reproduction form a dynamic balance that adjusts to the colony's needs and changes in its surroundings.

Knowing individual responsibilities in bee society is only one aspect of its understanding; another is the complex mechanisms of cooperation and communication that characterize hive dynamics. Bees use intricate chemical cues, such as pheromones secreted by the queen, to regulate activity and preserve colony unity as their primary means of communication. To ensure effective foraging and resource utilization, worker bees also engage in complex dances, such as the waggle dance, to transmit the location and quality of food sources to their nestmates.

Furthermore, every bee's life cycle and development correspond to its function inside the hive. Over their many years, queens continuously laid eggs to maintain the population of their colonies. On the other hand, workers usually live for a few weeks to a few months, depending on the colony's health, foraging risks, and seasonal demands. The drone caste has the lowest lifespan among the bee castes because they are seasonal and disposable for mating.

To sum up, understanding bee culture provides essential insights into the complexity and resiliency of these crucial pollinators. All bee members play a distinct part in the survival and success of the colony, from the queen's vital role in reproduction to the workers' adaptive behaviors and the drones' specialized purpose. In addition to ensuring the colony's survival, this complex social structure highlights how closely bees are linked to ecosystems and agriculture across the globe. Beekeepers and conservationists can work together to promote sustainable practices that support bee health and preserve their vital role in biodiversity and global food production by learning about and understanding bee society.

Bee communication (dances, pheromones)

Bees use two main methods for their remarkable efficiency and sophistication in communication: dances and pheromones. These communication techniques depend on coordinating hive activity, exchanging information about food sources, and preserving social cohesiveness within the colony.

The " waggle dance," used by honeybee foragers to notify nestmates about the location and quality of food supplies, is one of the most fascinating parts of bee communication. The waggle dance, which Austrian ethologist Karl von Frisch first observed in the 1920s, consists of a sequence of complex movements performed on the comb surface inside the hive. This dance is done by a foraging bee, which runs in a figure-eight pattern while wiggling its body and vibrating its belly. The direction of the food source with respect to the sun's position is indicated by the angle of the sway concerning the vertical comb, and the distance to the food source is correlated with the length of the waggle part. Worker bees use their ability to comprehend dances to locate and travel to far-off floral supplies precisely. This allows them to maximize foraging efficiency and guarantee that the colony's nutritional demands are met.

Bees use " pheromones," which are chemical signals, for comprehensive communication in addition to dances.

Pheromones are essential for controlling the behavior and structure of colonies. A range of pheromones are secreted by the queen bee, one of which is the queen mandibular pheromone (QMP), which prevents worker bees from developing ovaries and preserves the social hierarchy within the colony. The queen's reproductive status and presence are also indicated by QMP, which impacts worker behavior and colony health. The alarm pheromone, which bees emit in reaction to dangers or disruptions, is another critical pheromone. Other bees are alerted to possible danger by this chemical signal, which causes them to mobilize and engage in defensive actions like stinging to defend the hive.

Pheromones also aid in coordinating hive functions like swarming, foraging, and brood care. Produced by growing brood, brood pheromone aids in controlling worker duties inside the hive, fostering effective larval care and hive upkeep. Nasonov pheromone is left behind by foraging bees near the hive entrance to help orient newly emerging bees and direct returning foragers, facilitating smooth transitions and maintaining colony cohesion. Because they enable bees to react quickly to shifting environmental conditions and internal requirements, these chemical signals are essential to preserving cooperation and order within the hive.

The efficacy of pheromones and dances in bee communication highlights the flexibility and tenacity of social insect communities. Bees use these complex communication networks to manage risks, maximize resource use, and maintain colony productivity in ever-changing surroundings. Their capacity to use dances and pheromones to transmit exact information about food supplies and hive conditions improves their success in foraging, encourages successful reproduction, and develops group decision-making, which is crucial for colony survival.

Nevertheless, in contemporary environments, environmental stressors, including habitat loss, pesticide exposure, and climate change, threaten bee populations and alter their typical behaviors, making bee communication more complex. The sophisticated communication networks maintaining bee colonies and the world's food production depend heavily on conservation initiatives that protect bee habitats, minimize chemical inputs, and promote bee-friendly agricultural practices.

Finally, the intricacy and effectiveness of social insect societies are demonstrated by the communication methods used by bees, which include dances and pheromones. Bees can maintain social order in the hive through these systems, coordinate group actions, and relay important information. Researchers, beekeepers, and conservationists can create plans to improve bee health, protect pollination services, and increase the adaptability of bee populations in a world that is changing quickly by comprehending and valuing the subtleties of bee communication.

Foraging and pollination

Bees depend on foraging and pollination as vital components of their life cycle, necessary for agriculture, the larger environment, and bee survival. Foraging is an essential activity for bees, especially the hardworking worker bees, since it allows them to gather food, water, pollen, propolis, or plant resin, which is needed to keep the hive alive. Worker bees begin foraging when they leave the hive in pursuit of flowers that yield nectar, which is a sweet liquid that is their primary source of energy, and pollen, which is a protein-rich material that is essential for the growth of larvae and the nourishment of the hive. To forage efficiently, bees have evolved unique adaptations. These include an acute sense of smell that allows them to detect floral odors and ultraviolet vision that will enable them to recognize flower patterns that direct them to nectar and pollen sources.

One of the most important ecological services bees do is unintentionally aiding in pollination when they forage. Bees spread pollen grains between flowers when they visit them in search of nectar and pollen, necessary to reproduce many flowering plants. Fruits, seeds, and nuts are produced by a variety of crops and wild plants, and this process is essential to their growth. Bees are vital for preserving ecosystem stability and plant biodiversity because of their effectiveness and loyalty to particular plant species.

The levels of specialization in the foraging habits of different bee species vary. Certain bees, such as honeybees, are good generalist pollinators because of their broad diet and ability to feed on various plant types. Others, like some species of solitary bees and bumblebees, have specialized foraging habits that allow them to concentrate on particular plant families or types, which increases their effectiveness in pollinating those specific plants. Because different bee species have different feeding methods, robust pollination networks that maintain healthy plant ecosystems and agricultural productivity are supported.

The availability of floral resources and environmental factors directly impact the time and seasonality of foraging and pollination activities. The warmer months, when flowers blossom and provide an abundance of nectar and pollen, are when bees are busiest. Peak foraging seasons are spring and summer when growing and reproducing plants are at their best. To maximize honey output and colony health, beekeepers frequently arrange their hives to correspond with these seasons. This ensures that bees have access to plenty of forage.

In agricultural contexts, where controlled colonies greatly influence crop yields and quality, bee pollination is essential. Melons, apples, blueberries, almonds, and other crops rely primarily on bee pollination to produce fruit. Commercial beekeepers frequently move their hives to agricultural areas during bloom times to assist farmers and increase crop yield. This method of beekeeping, called migratory beekeeping, emphasizes the value of bees economically and their crucial role as collaborators in contemporary agriculture.

Nevertheless, several obstacles bees must overcome jeopardize their capacity to feed and pollinate crops efficiently. Several reasons, including illnesses, pesticide exposure, habitat loss, and climate change, are causing bee numbers to drop globally. To ensure that bees continue to play an essential role in pollination and the health of ecosystems, conservation measures that safeguard bee habitats, minimize the use of pesticides, encourage bee-friendly landscaping techniques, and maintain diversified pollinator populations are crucial.

In summary, bee biology and ecology are deeply entwined with foraging and pollination, underscoring the bees' essential function as ecosystem service providers and pollinators. While pollination services help biodiversity and global food production, their foraging activities maintain the health and productivity of the hive. Protecting bee populations, global agricultural systems and natural ecosystems depends on our ability to comprehend and assist bees in feeding and pollinating activities. We can ensure a future in which pollination thrives, bee populations increase, and ecosystems remain robust and diverse by cultivating a symbiotic interaction between bees and their surroundings.

Seasonal behaviors and hive activities

Bees' seasonal habits and hive operations are closely linked to the environment's natural cycles, showing natural adaptations that guarantee colony survival, reproduction, and productivity throughout the year. For beekeepers to efficiently manage hives and promote bee health, they must thoroughly understand these seasonal dynamics.

" Spring" marks the beginning of a busy time for bee colonies as they come out of their winter hibernation. Foraging activities recommence this season as temperatures rise and flower resources proliferate. Workers carefully gather pollen and nectar to feed developing larvae and restore food stores depleted by the winter. To quickly raise the colony's size in anticipation of the upcoming peak foraging season, queen bees inside the hive increase the pace at which they lay eggs. Bees labor nonstop to take advantage of favorable conditions and accumulate resources, with colony operations centered around brood rearing, comb construction, and hive upkeep.

" Summer" is the busiest time of year for bee activity, with active foraging, quick colony growth, and the building up of stored honey reserves. Bees increase their foraging activities, visiting various blooming plants to gather pollen and nectar. During this time, nectar is transformed by bees into honey, which is then kept in wax cells inside the hive. To keep the hive at the ideal temperature for the brood's development and the honey's ripening, worker bees ventilate the hive. During the summer, beekeepers inspect hives to evaluate the health of the colonies, manage hive space by adding supers (extra hive boxes), and ensure the bees have enough supplies to continue their activities.

" Autumn" changes hive dynamics as the temperature drops and the daylight hours shorten. As floral resources diminish, foraging activity steadily decreases, forcing bees to concentrate on winter survival strategies. To support brood development and protein reserves during winter, worker bees redirect their efforts toward gathering and storing pollen. Natural processes like the termination of brood rearing and the expulsion of older or sicker bees help colonies control their population growth. Activities in the hive include gathering honey storage, caulking entrances to control temperature, and preparing winter cluster formations, which are crucial for preserving heat and keeping the hive warm.

For bee colonies, " Winter" is a time of dormancy and conservation. In colder regions, bees hive in close quarters, keeping in touch and producing heat through movement to withstand subfreezing conditions. The queen, surrounded by worker bees at the cluster's center, is in charge. Bees use the honey they have stored as energy, causing their flying muscles to vibrate to provide warmth and ensure the colony survives until springtime arrives. To avoid starvation, hive weight is monitored, and honey stocks are evaluated throughout the winter months when fewer inspections of hives are conducted.

Bee colonies show extraordinary environmental resilience and adaptation throughout these seasonal behaviors and hive operations. Their coordinated reactions to seasonal cues, like temperature, duration of daylight, and the presence of flowers, guarantee effective use of resources, successful reproduction, and colony survival. To optimize honey production, design hive management methods, and maintain bee health through sustainable practices, beekeepers must be thoroughly aware of these seasonal changes.

But as the climate and environment change, bee colonies must contend with disease, pesticide exposure, and habitat loss. Bee populations must be protected, and their crucial role as pollinators must be supported through conservation initiatives that prioritize the preservation of a variety of floral resources, the reduction of chemical inputs, and the promotion of bee-friendly activities.

In conclusion, the seasonal behaviors and hive operations best illustrate the complex adaptation and social structure of bee colonies. Bees use accuracy and teamwork to handle seasonal shifts, from the hectic spring feeding to the careful winter conservation. Beekeepers may improve hive productivity, support bee health, and help preserve these vital pollinators and their priceless contributions to global agriculture and ecosystems by learning about and honoring these natural rhythms.

Signs of healthy hive behavior

A robust and efficiently operating bee colony can be determined by observing signs of healthy hive behavior, which also reflects the hive's general resilience, productivity, and health. These indicators help beekeepers determine the health of their colonies, spot possible problems early, and maintain ideal circumstances for the growth and productivity of their hives.

Active foraging and vigorous pollination are two main signs of a healthy hive. In colonies that are in good health, a large number of worker bees are frequently seen departing and returning to the hive with baskets full of vibrantly colored pollen and returning with nectar. Strong brood development and adequate nourishment are indicated by increased pollen gathering, while an abundance of nectar consumption influences honey output. Diverse foraging practices are crucial for balanced nutrition and the general health of the hive when bees are observed to return with different colored pollen.

A robust hive produces brood regularly and predictably all year long, which indicates the queen's ability to reproduce and the colony's health. During hive inspections, beekeepers evaluate brood patterns, looking for tightly closed cells that indicate developing pupae, open cells containing larvae at different stages of development, and eggs placed systematically. Consistent brood patterns among frames indicate a robust queen and sufficient nourishment, essential for preserving the population and robustness of the colony.

A healthy queen is essential to the longevity and productivity of the colony. Eggs placed inside cells in distinct patterns indicate a healthy queen. Resilient queens regularly deposit their eggs, arranged in the middle at the bottom of the chambers. A healthy hive environment is also marked by the lack of queen or emergency cells and worker bees showing signs of attentively tending to the queen and brood. Beekeepers also watch for a soft buzzing sound from within the hive, a sign of contentment and productivity.

When regular hive inspections are conducted, bees in healthy hives behave in a calm and collected manner and rarely react defensively. To control the temperature and humidity in the hive, bees must actively participate in hive chores such as feeding larvae, capping honey cells, and fanning. A well-balanced and productive colony is suggested by worker bees peacefully wandering throughout the hive without making loud or aggressive postures. On the other hand, aggressive, noisy, or defensive behaviors could be indicators of stress, illness, or environmental disturbances that need to be looked into further and treated.

Pollen, honey, and bee bread—a concoction of pollen and nectar that nourishes larvae—are among the resources a well-managed hive effectively arranges and stores. Robust hives preserve enough honey reserves to keep bees alive, like winter or bad weather, when little feed is available. Observing bees meticulously sealing honey cells, stacking brood frames in an organized manner, and exploiting comb efficiently reveals strong resource management and colony readiness for seasonal changes.

A vital sign of hive health, hygienic behavior reflects the capacity of the colony to keep itself clean and free from disease. Bees practice good hygiene by removing sick or dead broods, cleaning and uncapping cells for later use, and ventilating the hive to control humidity and temperature. Debris-free, spotless hive entrances and robust wax production are signs of a well-kept colony that is productive and healthy.

In summary, beekeepers may evaluate colony health, foster ideal conditions for bee development, and reduce hazards to hive health by identifying indicators of healthy hive behavior. Beekeepers support the resilience and sustainability of bee colonies by watching calming hive behavior, monitoring active foraging and pollination, evaluating brood generation and queen vitality, guaranteeing adequate resource storage, and encouraging hygienic practices. These metrics show bees' general well-being and productivity and the critical role that beekeepers play in maintaining healthy bee populations, which are necessary for pollination, biodiversity, and sustainable agriculture.

CHAPTER VI

Hive Management and Maintenance

Routine hive inspections

For beekeepers who want to guarantee their colonies' well-being, longevity, and production, hive management and upkeep—primarily through regular hive inspections—are essential procedures. Through systematic assessments of hive conditions, colony dynamics, and bee behavior, these inspections enable beekeepers to keep an eye on the health of their hives, gauge the availability of resources, and take preventive action when needed.

Throughout the beekeeping season, frequent hive inspections are essential, and the frequency of these inspections varies based on local forage availability, colony strength, and weather. In general, during the busy spring and summer foraging seasons, when hive activities are at their height, beekeepers conduct more frequent inspections. Inspection frequency may be reduced throughout the winter or during periods of bad weather to minimize disturbances and stress on the colony.

Regular hive inspections have several main goals, such as determining the size of the colony, tracking the growth of the brood, measuring the viability of the queen and her egg-laying habits, analyzing the honey storage, and spotting any indications of pests, illnesses, or anomalies in the hive. During inspections, beekeepers watch the general behavior of the bees, noting how they behave on frames, inside the hive, and at the entrance to determine the health and activity levels of the colony.

Before conducting inspections, beekeepers gather equipment such as a smoker to soothe bees, a hive tool to break apart frames, and protective clothing, such as a beekeeping suit, gloves, and veil to reduce stings. Sufficient preparation allows for comprehensive and adequate hive inspections while guaranteeing the bees' and beekeepers' safety. Furthermore, beekeepers select the best times of day to conduct inspections; these are ideally mildly tempered mid-morning or early afternoon hours when bees are most likely to be at ease and working in their hives.

Beekeepers approach the hive quietly during a routine examination, using light smoke to calm the bees and lessen their defensive reactions. The hive cover and frames are carefully removed, and each frame is inspected individually for evidence of a healthy brood, sufficient supplies of pollen and honey, and the queen's presence. To determine the fertility and health of the queen and the colony, beekeepers follow trends in the development of eggs, larvae, and pupae at different stages of life. They look for queen cells, meaning swarming preparations are underway or an elderly queen needs to be replaced.

Good hive management involves keeping thorough records of hive inspections, documenting observations on the colony's health, the state of the queen, the presence of pests and diseases, and hive conditions all year. These logs are invaluable for monitoring colony development, spotting patterns or problems, and guiding management choices on extra feeding, hive growth, and pest and disease control procedures.

Beekeepers can apply several management tactics to boost their hives' productivity and health based on the inspection results. Adding or removing hive boxes (supers) to accommodate colony growth or reduce space during winter, treating for pests or diseases using approved methods that minimize harm to bees and hive integrity, and feeding supplemental sugar syrup or pollen patties to bees during periods of nectar dearth are a few examples of how to do this.

It takes continual education and learning about bee biology, behavior, and best practices for beekeeping to manage hives successfully. Beekeepers maintain up-to-date knowledge on developments in beekeeping methods, disease and pest control tactics, and environmental elements that impact bee health. Participating in workshops, becoming a member of beekeeping associations, and building relationships with seasoned beekeepers offer chances for the information and skill growth necessary for efficient hive management. Regular hive inspections are crucial to ethical beekeeping because they allow keepers to monitor colony health, evaluate hive conditions, and take prompt action to promote bee welfare and increase hive productivity. Beekeepers support the critical function of bees in pollination, biodiversity, and agricultural sustainability while fostering the resilience and sustainability of bee populations through careful inspections, suitable instruments and protective gear, and meticulous record-keeping. Beekeepers are vital to preserving these essential pollinators and their contributions to ecosystems worldwide because of their commitment to hive management techniques that put bee health and well-being first.

Feeding your bees

A vital component of ethical beekeeping is feeding the colonies to ensure their health, survival, and productivity —especially in times of nectar scarcity, lousy weather, or establishing new colonies. When floral resources are scarce, beekeepers supplement natural food with additional nutrition to keep bees alive. Knowing when and how to feed bees to promote hive development and reduce any dangers to the colony's welfare is crucial.

Beekeepers frequently utilize two forms of supplemental feeding: sugar syrup and pollen substitutes. " Sugar syrup" offers the carbohydrates needed for hive maintenance and energy generation, replacing nectar. It usually consists of a blend of water and granulated sugar, with the proportions changing according to the demands of the colony and the seasons. Depending on whether the purpose is to promote brood rearing or get ready for winter, sugar syrup can have a thin (1:1 sugar-to-water ratio) or thick (2:1 sugar-to-water ratio) consistency. In times when natural pollen sources are limited, beekeepers also employ "pollen substitutes" to complement natural pollen sources, offering vital proteins, lipids, vitamins, and minerals required for larval growth and colony health.

Beekeepers use the strength of their colonies, the season, and the availability of fodder to determine when to feed. Feeding is essential in the spring and early summer when colonies proliferate and need lots of food to raise their young. Feeding during these times promotes the colony's growth and increases brood production, strengthening the hive. To guarantee that colonies have enough stocks of honey and pollen for winter survival when forage becomes scarce, beekeepers may also give supplemental feeds in the late summer and early fall.

Beekeepers use various techniques to provide supplemental feeds based on the hive's design and the feeding goals. The most popular approach is to put feeders inside or close to the hive so that bees can readily get the food. Popular options include frame feeders, entrance feeders, and top feeders; all provide benefits like accessibility, pest prevention, and simplicity of refilling. Entrance feeders are placed near hive entrances for easy access, whereas frame feeders fit straight into the hive to reduce disruption. Large feed reservoirs are provided by top feeders, positioned above the inner cover and lessen the chance of robbery by other bees.

Feeding is necessary to keep colonies strong, beekeepers must carefully monitor feed consumption and hive conditions to avoid feeding their colonies too much or too little. Overfeeding can cause stored feed fermentation, hive moisture issues, and heightened vulnerability to pests and illnesses. Conversely, underfeeding can lead to hunger, weakening colonies, and lower winter survival rates. Based on observed consumption rates, the dynamics of the hive population, and meteorological variables that impact the availability of forage, beekeepers modify feeding schedules and quantities.

While supplemental feeding is essential to beekeeping methods, especially in areas with variable weather patterns or limited floral resources, natural foraging is recommended for bee nutrition and colony health. To preserve the health and resilience of their colonies, beekeepers establish a balance between encouraging the bees' natural foraging habits and offering sufficient supplemental feeds as needed. Sustainable beekeeping techniques ensure bees play a crucial part in food production and agricultural sustainability by promoting biodiversity, ecosystem health, and efficient pollination.

Beekeepers may need to implement emergency feeding protocols to save remaining colonies from hunger in the event of unanticipated food shortages or emergencies, such as late-season colony losses or abrupt weather changes. To ensure bee survival during difficult times, emergency feeding entails giving concentrated feed solutions and continuously monitoring hive conditions.

In summary, providing food for bees is an essential part of ethical beekeeping methods, ensuring colonies' health, survival, and productivity all through the beekeeping season. Beekeepers may effectively manage hive nutrition, reduce hazards, and promote sustainable beekeeping operations by being thoroughly aware of the many types of feeds, proper feeding times, methods, considerations, and monitoring measures. Beekeepers ensure that bees play a vital part in ecosystems and global food security by conserving pollinators and promoting agricultural sustainability through their careful feeding practices to maintain robust and resilient bee colonies.

Managing hive pests and diseases

An essential part of beekeeping is controlling illnesses and pests in the hive, which is necessary to preserve the colony's longevity, health, and production. Pests, diseases, and environmental stressors can weaken colonies, interfere with hive operations, and endanger bee numbers, among other issues beekeepers face. To reduce risks and foster resilient bee colonies, proactive monitoring, prompt interventions, and integrated pest management (IPM) techniques are all essential components of effective pest and disease control strategies.

The " Varroa destructor mite" is a parasite mite that feeds on adult bees and their offspring, weakening the bees and spreading viruses that can destroy entire colonies. It is one of the most frequent pests that damage beehives. Using sticky boards, alcohol washes, or powdered sugar shakes, beekeepers frequently check the amounts of Varroa mites to gauge infection levels and establish treatment thresholds. When infestations surpass thresholds, integrated pest management strategies for Varroa mites may involve chemical treatments using approved miticides, mechanical controls like screened bottom boards, and cultural methods like brood disruption.

Numerous diseases that can affect adult bee health, brood development, and colony viability are caused by bacteria, viruses, fungi, and protozoa, and they can affect bee colonies. Common bee diseases include " American foulbrood (AFB)," a bacterial disease that affects bee larvae, and "Nosema" infections, caused by parasitic microsporidians that interfere with bee digestion and absorption of nutrients. Early detection and diagnosis of infections are essential to apply timely therapies, such as antibiotics for AFB or fumagillin treatments for Nosema, and to adhere to veterinary and regulatory requirements to reduce resistance and protect colony health.

Regular hive inspections are essential for managing pests and diseases because they let beekeepers see how bees behave, examine brood patterns, and spot disease symptoms or indications of infestation. Beekeepers might gather samples for laboratory examination to validate disease diagnosis and inform the best course of action. Monitoring instruments, including thermal imaging, hive scales, and molecular diagnostics, improve surveillance and allow early identification of diseases and pests before they become more severe and affect colony health.

Biosecurity Precautions: Biosecurity protocols must be implemented to stop diseases and pests from entering and spreading across apiaries. Beekeepers maintain proper hygiene by limiting access to the hive to authorized personnel only, sterilizing or replacing hive tools after each use, and washing and sanitizing equipment after each use. Maintaining biosecurity standards to protect bee health and apiary integrity involves quarantining recently purchased bees or equipment, exercising prudent colony management, and minimizing the chance of introducing infections and pests.

IPM, or integrated pest management: By combining several tactics, an IPM method can effectively manage diseases and pests while reducing the need for chemical treatments. Keeping robust colonies, encouraging genetic diversity, and giving bees enough food to boost their immunity and resilience are examples of cultural behaviors that are part of IPM tactics. In order to organically and sustainably lower pest populations, physical measures such as screened bottom boards or hive beetle traps work in conjunction with biological controls like beneficial insects that feed on hive pests.

To improve the management of pests and diseases in beekeeping, ongoing study and instruction are essential. Beekeepers remain informed by working with researchers and extension specialists, attending workshops and seminars, and staying up to date on emerging threats, pest resistance issues, and innovative management strategies. Engaging in citizen science projects and contributing to disease surveillance programs facilitates gathering data and group endeavors to safeguard bee populations against worldwide hazards.

Beekeepers follow regional, governmental, and global laws controlling pests and diseases in beekeeping. Compliance involves reporting disease outbreaks, registering apiaries, and adhering to advised treatment regimens to stop the spread of diseases and pests between areas. Working with veterinary specialists, regulatory bodies, and other beekeepers encourages prudent apiary management and bee health stewardship.

In summary, controlling hive pests and illnesses is essential to sustainable beekeeping methods to maintain bee colonies' resilience, productivity, and well-being. Beekeepers protect bee populations and sustain their essential role as pollinators through proactive monitoring, integrated pest control tactics, biosecurity measures, and continuous education and research. By working together and using ethical management techniques, beekeepers support agricultural sustainability, biodiversity preservation, and global food security while acknowledging bees' vital role in ecosystems and human welfare worldwide.

Swarming: prevention and control

Swarming is a natural occurrence in bee colonies, indicating the need for reproduction and colony growth. Swarming is a way for colonies to multiply. Still, it can also cause problems for nearby communities, interrupt the production of honey, and result in the loss of valuable workers and queens. Beekeepers utilize diverse tactics to avert and manage swarming, guaranteeing the well-being and efficiency of their colonies.

When colony densities are at their highest and resources are plentiful, swarming usually occurs in spring and early summer. Overcrowding in the hive, which causes the colony to split up and start new colonies elsewhere, is the leading cause of swarming. About half of the worker bees leave the hive with the old queen to form a swarm, and a new queen is raised to replace the old one during the swarming process. A swarm is defined as a dense group of bees that, while scout bees look for a suitable new hive location, frequently cling to adjacent objects or trees.

Proactive hive management that addresses overpopulation and promotes colony health is the first step towards preventing swarming. Beekeepers may track their colonies' expansion, evaluate their hives' capacity to store honey, and spot early warning indicators of impending swarms, including the presence of queen cells, by conducting routine hive inspections. Congestion and the chance of a swarm can be lessened by adding supers, or extra hive boxes, to the hive during high population development. Stabilizing the colony and lowering swarming tendencies can also be achieved by controlling nutrition for the colony by providing extra food during nectar scarcity and making sure the hive is maintained and has enough air.

Beekeepers can step in to regulate swarming behavior and lessen its adverse effects on colony health and productivity when they notice warning signals of swarming. "Splitting" the colony physically into two or more independent hives, each with its queen and complement of bees, is one efficient method. Beekeepers preserve queens and bees by artificially forming swarms, which permits colonies to grow organically. Another strategy is " queen management," in which beekeepers raise young queens from selected stock to replace aging or underperforming queens, thus lowering swarm impulses and increasing productivity in the hive.

Stunting the development of queen cells is another way to stop swarming. Beekeepers temporarily confine the queen in a queen excluder inside the hive, limiting her movement and lowering her egg-laying pace to remove queen cells or suppress the swarming instinct. By focusing the colony's attention on producing honey and raising offspring instead of preparing for a swarm, this method preserves the stability and productivity of the hive.

To create successful swarming prevention and control plans, beekeepers can use the information and tools beekeeping associations, extension services, and knowledgeable mentors offer. Beekeepers can effectively anticipate and control swarming episodes by learning about the behaviors and triggers of swarming, comprehending the local environmental elements influencing swarming inclinations, and applying best practices in hive management.

Social interaction with neighbors and local communities is essential to responsible swarm management. Building goodwill and reducing worries about bee presence are achieved by teaching neighbors about beekeeping, emphasizing the role swarming plays in colony reproduction, and providing aid with swarm recovery and relocation. To ensure the welfare of the bees and the safety of the Community, beekeepers work in conjunction with local beekeeping groups and pest management experts to securely catch and transport swarms to acceptable hive locations.

Environmental factors such as pesticide exposure, habitat loss, and climate change can impact Swarming behaviors and colony health. Beekeepers support sustainable land management techniques, habitat restoration initiatives, and decreased pesticide use to maintain bee populations and lessen pressures contributing to swarming tendencies. Beekeepers support pollinator protection laws and bee-friendly habitats, which help preserve bee biodiversity and the health of ecosystems.

In conclusion, proactive hive management, an awareness of bee behavior, and successful intervention techniques are necessary to avoid and manage swarming in bee colonies. Beekeepers enhance colony health, limit disturbances caused by swarming, and contribute to sustainable beekeeping practices through preventive measures, swarm control techniques, education, and community engagement. Beekeepers preserve the vital function of bees as pollinators, promoting biodiversity, agricultural sustainability, and global food security through conscientious management and cooperation.

Harvesting honey and other hive products

A satisfying way to end the beekeeping season is to harvest Honey and other products from the hive. This shows how hard beekeepers work to maintain healthy colonies and encourage sustainable methods. In addition to Honey, beekeepers can extract other products from their hives, like Propolis, beeswax, royal jelly, and pollen, each with its uses and advantages.

The result of the bees' foraging activities and the beekeeper's management techniques is collecting Honey. When hive frames are capped, a sign that Honey is mature and ready to be extracted, beekeepers usually harvest Honey. Honey supers, or the boxes that hold the Honey, are first taken out of the hive and moved to a particular honey house or processing location. Carelessly unscrew the frames using a heated knife or uncapping fork to reveal the honey-filled cells. After that, extractors are utilized to spin the frames so that Honey may be extracted without destroying the wax comb by employing centrifugal force. Before the extracted Honey is kept in hygienic, food-grade containers for consumption or sale, it is filtered to eliminate contaminants like air bubbles and wax particles.

Another priceless hive product that beekeepers extract is beeswax, which is highly valued for its versatility in woodworking, candle making, and cosmetics. Beekeepers gather leftover wax from hive components and cappings taken out of honey frames during the honey extraction. Clean, golden beeswax blocks or pellets are produced by melting and sifting beeswax to eliminate debris and impurities, making it ready for usage. Reusing beeswax after Honey is extracted minimizes waste and makes the most use of available resources in the apiary.

Propolis, sometimes known as "bee glue," is a resinous material that bees gather from sap flows and tree buds. It is prized for its antibacterial qualities and potential uses in medicine. Propolis can be collected by beekeepers either by scraping it off hive surfaces or by employing propolis traps to encourage bees to deposit it in collection trays. After being processed to remove beeswax and debris, Propolis yields a black, aromatic resin for its antibacterial and anti-inflammatory qualities in health supplements, natural medicines, and cosmetics.

Worker bees create royal jelly, a nutrient-rich fluid supplied to queen and larval bees to aid their growth and development. Using specialized grafting techniques, beekeepers transplant immature larvae to colonies that produce royal jelly and are outfitted with royal jelly cups to gather royal jelly. Harvested royal jelly, prized in health supplements and cosmetics for its alleged health advantages and skin-rejuvenating qualities, is handled carefully and refrigerated to maintain its freshness and nutritional integrity.

By providing a variety of food crucial for the upbringing of young bees and the colony's well-being, pollen gathering enhances beekeeping methods. Beekeepers place Pollen traps at hive entrances to allow bees to enter and gather pollen pellets that land in collection drawers before being kept or marketed as a natural food supplement high in proteins, vitamins, and minerals; collected pollen is dried and processed to remove excess moisture and retain nutritional content.

To preserve the health of bee colonies and advance sustainable beekeeping, beekeepers prioritize ethical harvesting methods. Among the practices are limiting disruption during harvesting to minimize stress on bees, using natural and environmentally friendly processing processes for hive products, and providing bees with sufficient honey reserves over the winter to maintain colony health. Beekeepers maintain moral norms that promote bee welfare and environmental stewardship by managing hives responsibly and honoring natural bee behaviors.

Beekeepers gain from harvesting hive products, supporting local economies and communal sustainability. Beekeepers market their Honey, beeswax products, Propolis, royal jelly, and pollen at regional farmers' markets, specialty shops, and online to showcase the caliber and variety of their apiary goods. Beekeepers help biodiversity conservation, increase public understanding of the use of pollinators in agriculture, and improve food security and ecosystem health by marketing their products.

To sum up, gathering Honey and other products from the hive is the ultimate goal of beekeeping, demonstrating the benefits of maintaining a healthy colony and practicing sustainable hive management. Beekeepers are essential to preserving bee populations, promoting agricultural sustainability, and disseminating the advantages of natural hive products to communities across the globe. They accomplish these goals through engaging in local marketplaces, maintaining ethical standards, and encouraging appropriate harvesting procedures. Beekeepers support bees' vital role as pollinators and advocate for sustainable methods in beekeeping and beyond through their commitment to bee welfare and environmental care.

CHAPTER VII

Troubleshooting Common Problems

Identifying common issues (disease, pests, environmental stress)

To troubleshoot typical beekeeping challenges, one must first identify and handle a variety of concerns that may impact the production and health of bee colonies. To reduce risks and promote the welfare of bees, beekeepers need to be on the lookout for diseases, pests, and environmental stresses. They must also be alert in monitoring hive conditions and identifying warning indicators.

Managing illnesses that can harm colonies and influence bee populations is one of the main difficulties beekeepers encounter. "American foulbrood (AFB)" is a highly contagious illness that affects bee brood and is characterized by foul-smelling, ropey larvae and darkened cell walls. The bacteria cause "Paenibacillus larvae". Beekeepers visually inspect brood frames to diagnose AFB; laboratory analysis confirms the diagnosis. Treatment usually entails eliminating affected colonies and putting biosecurity measures in place to stop the disease from spreading. Another prevalent ailment brought on by microsporidian parasites that impair bee vigor and digestion is " Nosema." To control Nosema outbreaks and promote colony recovery, beekeepers treat their colonies with fumagillin and watch for symptoms like diarrhea and diminished colony strength.

In beekeeping, controlling pests is essential to avoiding colony infestations and minimizing hive disturbances. One of the most destructive pests is still the "Varroa destructor mite," which feeds on bee hemolymph and spreads viruses that damage bees and cause colony loss. Using alcohol washes or sticky boards, beekeepers routinely check the amount of Varroa mites in their hives. They then use integrated pest management techniques to control the mite population. Biological controls like selective breeding for mite-resistant bee stock, cultural techniques like removing drone broods to stop mite reproduction, and chemical treatments using licensed miticides are some of the methods used.

Environmental factors contributing to bee health issues and colony vulnerabilities include habitat loss, pesticide exposure, climatic variations, and nutritional inadequacies. In addition to promoting sustainable land management techniques and pesticide reduction initiatives, beekeepers evaluate environmental factors affecting the availability of feed, water sources, and pesticide exposure in the area. Colony resilience is enhanced, and stress-related health problems are reduced by supplementing nutrition during nectar scarcity periods and ensuring the hive is ventilated and insulated from harsh weather.

When a hive loses its queen, the generation of brood is decreased, and the colony's strength declines. Beekeepers must act quickly to address queen-related issues, as they can upset colony dynamics and productivity. To reestablish hive stability, beekeepers identify queenlessness by observing changes in worker behavior and a decrease in egg-laying activity. They then introduce a new queen or brood frame holding eggs for queen raising. " Queen Supersedure": When queen cells are found in the hive, the bees replace an aged or underperforming queen. If beekeepers want to preserve genetic diversity and productivity, they can either wait for natural supersedure or act by requiring the colony with a new queen.

"Crowding:" Swarming is a standard way for colonies to reproduce, but it can also reduce honey yield and cause the loss of valuable queens and worker bees. Beekeepers use swarm prevention strategies, including colony splitting, queen management, and giving plenty of hive space at population peaks. They also watch for indicators of impending swarms, such as queen cell production and packed hive space. Swarms can be contained and relocated to bait hives or new hive boxes to reduce colony losses and encourage regulated colony growth.

To effectively troubleshoot beekeeping, community involvement, information exchange, and resource accessibility are necessary. Beekeepers engage in local beekeeping associations, workshops, and online discussion boards to share knowledge, get guidance on problem-solving techniques, and remain up to date on best practices and new challenges. Working with researchers, veterinary specialists, and extension services promotes continuous learning and research initiatives, improving beekeeper competencies in disease identification, pest control, and environmental stewardship.

In conclusion, to protect bee health and advance sustainable hive practices, resolving typical issues in beekeeping calls for proactive management, astute observation, and well-informed decision-making. Beekeepers assist agricultural pollination, maintain the vital function that bees play in ecosystems around the world, and help to strengthen the resilience of bee populations by quickly and effectively diagnosing and resolving diseases, pests, environmental stressors, queen concerns, and swarming tendencies. Beekeepers support the long-term sustainability of beekeeping techniques for future generations by upholding ethical standards, engaging the community, learning continuously, and practicing responsible stewardship.

Solutions and treatments

In beekeeping, "solutions and treatments" refer to various tactics to deal with typical problems such as illnesses, pests, environmental stressors, and difficulties managing hives. Beekeepers employ ethical treatments, integrated pest management (IPM) techniques, and preventive measures to support colony health, production, and sustainability.

Vigilant surveillance and early diagnosis of bee diseases such as " American foulbrood (AFB)" and "Nosema" are essential for effective disease treatment. Beekeepers use diagnostic instruments such as microscopy or laboratory testing to make an accurate diagnosis after doing routine hive inspections to spot illness symptoms, such as foul-smelling brood and dysentery. Various treatment plans are available; nevertheless, to stop the spread of AFB, beekeepers must frequently burn contaminated colonies and maintain high hygiene standards. Fumagillin or other licensed drugs are given as part of the "Nosema" therapy regimen to lessen the effects of parasites and aid in colony recovery. Maintaining robust colony genetics, implementing biosecurity protocols, and reducing stressors heightening susceptibility to illness are all integral parts of integrated disease management.

Controlling pests like the " Varroa destructor mite" is essential for the longevity and output of colonies. Because varroa mites feed on bee hemolymph and spread viruses, they harm bees. Therefore, regular mite monitoring using tools like sticky boards or alcohol washes is necessary. Integrated pest management systems aim to reduce chemical residues in hive products while mitigating mite infestations through cultural, biological, and chemical controls. To maintain bee health and hive integrity, beekeepers can use natural predators such as specific mite-eating beetles, apply miticides sparingly and by label recommendations, or use selective breeding to produce bee stock resistant to mites.

It takes teamwork and responsible stewardship to address environmental stresses like habitat loss, pesticide exposure, and nutritional deficits. Beekeepers support agricultural methods that are bee-friendly, reducing the use of pesticides and offering a variety of fodder sources that are vital to bee nutrition. In drought or lack of nectar, supplemental food keeps colonies solid and resilient, and well-ventilated, well-insulated hives shield bees from harsh weather. Beekeepers support sustainable beekeeping practices and ecosystem health by supporting measures to reduce pesticide use and conserve habitat.

Problems with the queen, such as "queenless ness" or " poor queen performance," affect the dynamics and output of the colony. When hive conditions allow, beekeepers might allow natural supersedure or introduce new queens raised from selected stock to handle queen concerns. To maintain the genetic diversity and health of the colony, requeening entails carefully introducing a new queen or queen cells to replace an elderly or failing queen. Beekeepers can respond quickly to maintain hive productivity by observing worker behavior and egg-laying patterns, which provide insight into the queen's performance.

Although swarming is a normal behavior for bees, it can lower honey production and result in colony losses if not controlled. Swarm prevention strategies beekeepers use include "splitting colonies," giving hives enough room and controlling queen cells to deflect swarming impulses. Beekeepers can manage colony expansion under controlled settings, minimize interruption to hive activities, and preserve valuable bees by capturing and moving swarms to new hive boxes or bait hives.

In beekeeping, community involvement, knowledge exchange, and continuous education contribute to more effective answers and treatments. Beekeepers engage in local beekeeping associations, workshops, and online discussion boards to share knowledge, get guidance on best practices, and remain up to date on new issues and creative solutions. By working with researchers, veterinary specialists, and extension services, beekeepers can enhance their knowledge of disease management, pest control, and sustainable hive practices through ongoing learning and research projects.

Conclusively, the successful use of solutions and treatments in beekeeping necessitates a comprehensive strategy that incorporates ethical therapies, proactive management, and ongoing education. Beekeepers keep bee populations healthy and resilient, aid in agricultural pollination, and assist biodiversity conservation by using informed tactics and responsible practices to address illnesses, pests, environmental stressors, and hive management difficulties. Beekeepers are crucial in preserving bees' vital roles in ecosystems and worldwide food security through their commitment to bee welfare, community involvement, and ecological management.

When to seek professional help

In beekeeping, knowing when to call in a professional is essential to keeping colonies healthy, dealing with complex problems, and guaranteeing good management techniques. Even if many beekeeping duties may be completed on your own, some circumstances call for the knowledge of entomologists, veterinarians, or seasoned beekeepers.

Seeking advice from a qualified entomologist or bee disease specialist is crucial for beekeepers who experience unusual symptoms or suspect disease outbreaks like "American foulbrood (AFB)" or " European foulbrood (EFB)." Through laboratory analysis, these professionals can accurately diagnose conditions and offer treatment plans that adhere to regional laws. Without compromising bee health or hive integrity, professional advice provides effective illness control, limits disease transfer to neighboring hives, and supports colony recovery.

To stop colony collapse, controlling pests such as the " Varroa destructor mite" necessitates careful monitoring and action. If beekeepers see elevated levels of mites or have difficulties managing mite populations using integrated pest management strategies, consulting with seasoned beekeepers or agricultural extension specialists is recommended. These experts can suggest alternate control methods, provide information on cutting-edge treatment choices, or grant access to specialist tools for accurate mite monitoring and management.

To preserve hive productivity and genetic diversity, professional intervention may be required for queen-related issues such as " queen failure," "supersedure," or "swarm control." Seeking advice from experienced bee breeders or beekeeping mentors is beneficial for beekeepers who are having trouble with their queens or who require help with requeening methods. These professionals can evaluate the queen's health, advise on strategies for introducing queens, or give access to superior queen bees bred for maximum performance and resistance to disease.

When faced with environmental stresses like "pesticide exposure," "habitat loss," or "forage scarcity," beekeepers should consult with conservationists, ecological scientists, or agricultural professionals for guidance. Environmental assessments, suggestions for sustainable land management techniques, and lobbying for pollinator-friendly legislation that promotes bee health and habitat preservation can all be made by experts. Working with specialists guarantees proactive approaches and well-informed decision-making to reduce environmental effects on bee populations and the sustainability of apiaries.

Complying with local, national, and international regulations about beekeeping methods necessitates familiarity with industry standards and legislative requirements. When navigating legal issues like "hive registration," "disease reporting," or "pesticide use," beekeepers seek guidance from regulatory bodies, extension agents, or agricultural law experts. Expert advice guarantees adherence to legal requirements, reduces legal risks, and encourages ethical beekeeping methods that respect environmental stewardship and biosecurity.

" Emergency Situations:" Beekeepers should put safety first and seek out expert help right away in cases of "hive collapse," " severe weather damage," or "bee stings" that require medical attention. Making the necessary arrangements with emergency services, medical professionals, or seasoned beekeepers versed in emergency response techniques guarantees rapid medical attention for accidents connected to bees, appropriate hive recovery techniques, and timely intervention. In an emergency, expert help ensures the safety of beekeepers and facilitates a quick recovery and restoration of hive health and functionality.

" Ongoing Education and Assistance:" Participating in continuous education, mentorship initiatives, and cooperative networks within the beekeeping community improves beekeepers' abilities, know-how, and self-assurance in handling a variety of difficulties. Attending workshops, seminars, and online forums allows for sharing knowledge, access to expert counsel on complex beekeeping challenges, and troubleshooting tips. Beekeepers enhance their ability to make well-informed decisions, adjust to changing circumstances, and sustainably manage bee colonies for maximum well-being and productivity by cultivating a culture of ongoing learning and assistance.

To sum up, beekeepers who know when to seek expert assistance are better equipped to handle intricate situations, preserve the health of their colonies, and adhere to industry best practices for apiary management. Beekeepers ensure responsible management of bee populations, support environmental sustainability and contribute to global efforts in pollinator conservation and agricultural resilience by utilizing the knowledge of entomologists, bee disease specialists, agricultural professionals, and regulatory authorities. Beekeepers are critical in preserving bees' significant contributions to food security, ecosystems, and global human well-being through cooperative efforts and proactive engagement with professional resources.

Seasonal challenges and how to handle them

Seasonal beekeeping problems require strategic planning, proactive management, and adaptive reactions to ensure colony health and productivity throughout the year. Maintaining effective apiary operations across various climates and regions requires understanding these seasonal changes and applying appropriate handling practices.

" Spring:" In bee colonies, springtime signals a time of expansion and activity characterized by increased brood generation, foraging, and colony growth. But this season also has drawbacks, like " swarming," which occurs when colonies split off to increase and may lower honey yield. To prevent swarming, beekeepers should regularly inspect their hives to check the number of colonies, the generation of queen cells, and the available space. Adding more supers to allow for colony expansion and using swarm avoidance methods like " splitting" or "queen management" can help reduce the likelihood of swarming and maintain hive stability during this crucial time.

Summer:" Beekeepers can produce a lot of honey throughout the summer, but concerns include "heat stress" and "water scarcity." Bee colonies may experience stress from high temperatures, reducing foraging efficiency and even overheating in hives. Beekeepers mitigate heat stress by ensuring hives have enough ventilation and shade, situating colonies in areas with shade where feasible, and keeping water supplies close to the apiary to avoid dehydration. Despite the obstacles of summer, beekeepers may monitor honey output and efficiently manage hive conditions by keeping an eye on colony productivity and behavior.

Autumn:" For bee colonies preparing for winter survival, fall marks a temperature shift and sunshine length. "Management of varroa mites," "nutritional deficiencies," and "winterization of colonies" are significant issues. In the fall, beekeepers double their efforts to treat and check for varroa mites to reduce their populations before winter clustering. Evaluating beehive reserves and adding " sugar syrup" or "pollen patties" to the mix guarantees that colonies have enough food for winter survival and brood rearing. In addition, " winter wraps" or "ventilation boards" can insulate hives and reduce entrances, aiding bee heat conservation and hive warmth throughout the colder months.

Cold:" As colonies enter a decreased activity and resource shortage phase, winter presents unique problems. If not correctly handled, freezing temperatures and extended isolation can strain bee health, resulting in "starvation," "condensation," and "queen losses." Prioritizing hive insulation and "winter feeding" is how beekeepers keep their colonies going during hard times. Supplemental winter feed in the form of "fondant or candy boards" keeps hives dry and gives bees access to vital carbohydrates. On moderate winter days, beekeepers can monitor hive weight and periodically visit the colony to determine health, identify emergent problems, and give interventions to encourage successful overwintering.

Managing seasonal beekeeping difficulties necessitates localized temperature adjustments, forage availability, and environmental factors. Beekeepers throughout various regions may encounter unique obstacles like "drought," "pollution," or "crop spraying," which calls for customized approaches to pest control, habitat improvement, and hive defense. Working with local environmental authorities, extension services, and beekeeping associations gives beekeepers access to knowledge, resources, and support networks unique to their area, improving their ability to manage seasonal fluctuations and maintain the health of their apiaries.

Effective management of seasonal problems requires a commitment to lifelong learning and readiness. To stay current on new developments, industry best practices, and creative solutions, beekeepers participate in peer-to-peer knowledge sharing, workshops, and continuing education. Beekeepers can enhance their capacity to protect colony health, maximize honey yield, and advocate for sustainable beekeeping methods throughout the year by proactively managing their colonies, predicting seasonal variations, and modifying their strategies to suit local conditions.

To sum up, managing seasonal obstacles in beekeeping requires resilience, foresight, and adaptive management techniques to maintain the health and output of bee colonies. Beekeepers support pollinator protection, agricultural sustainability, and ecosystem health by appreciating seasonal dynamics, using proper handling practices, and utilizing local adaptations. Beekeepers preserve bees' essential function as pollinators and biodiversity protectors by working together, acquiring new skills regularly, and being proactive in observing seasonal subtleties. This ensures a sustainable future for bee populations worldwide and the welfare of humankind.

CHAPTER VIII

The Science of Beekeeping

The biology of honey production

The complex biology and behaviors of honeybees are extensively studied in beekeeping, especially the science underlying honey production. The biology of honey production, which results from the bees' extraordinary physiological adaptations and interactions with their surroundings, must be understood to comprehend the process thoroughly.

The way worker bees forage is where the biology of honey production starts. These hardworking insects travel many miles from their hives to find pollen and nectar sources. Primarily, foraging bees gather nectar from blooming plants, the starting point for making honey. Bees collect nectar using their long, tube-shaped proboscis to scrape the sweet liquid from flowers and temporarily store it in their honey-filled stomachs.

A forager bee brings its bounty of nectar back to the hive and regurgitates it for the eager attention of the house bees. The process of turning raw nectar into honey begins with this exchange. After consuming the nectar, the house bees continually regurgitate and consume it again, adding salivary gland enzymes to help break down complex sugars into simpler ones like fructose and glucose. Nectar undergoes an enzymatic procedure that improves chemical composition and lowers water content—two essential processes in honey production.

After gathering nectar, house bees carefully place it into the honeycomb's hexagon-shaped wax cells. These cells, made from beeswax generated by younger worker bees, offer mature honey a safe and well-organized storage space. The bees furiously flap their wings to encourage air circulation and further lower moisture levels as they fill each cell with partially dehydrated nectar, guaranteeing the honey's quality and storage.

The ripening and capping of the honeycomb cells are the last steps in producing honey. When the honey achieves its ideal moisture content of approximately 17–18%, house bees utilize beeswax cappings to seal the filled cells. The honey's long-term stability and nutritional integrity are protected by this protective coating, which also inhibits moisture absorption and microbial contamination. Honey continues to develop and ripen inside the sealed cells, gaining more flavor, viscosity, and nutritional density as it does so.

The biology of honey production is significantly impacted by environmental elements such as hive management techniques, floral diversity, and weather patterns. Bees flourish in habitats rich in a diversity of blooming plants that provide a range of nectar sources, which affects honey's nutritional makeup and flavor profiles. Strategically placing hives in floral resource-rich areas and keeping an eye on seasonal bloom cycles allow beekeepers to maximize honey yield and quality.

To maintain hive health and optimize productivity, honey harvesting requires precise timing and approaches. Beekeepers remove honey-filled frames from hives and take them to extraction facilities using smoke to calm the bees. In this instance, beeswax is preserved as honeycomb cells are extracted using centrifugal force after being uncapped. Before the extracted honey is bottled and stored for consumption and distribution, contaminants are removed through filtration.

Scientific developments in beekeeping contribute to our ongoing study of bee biology and honey production. Research on the diet, behavior, and genetics of bees helps to manage hives sustainably and guides breeding efforts for disease-resistant bee stock. Scientific advancements in hive design, pest management, and honey processing technology boost bee health and ecosystem resilience while increasing beekeepers' productivity, quality, and efficiency.

In summary, the science of honey production embodies the mutually beneficial relationship between the biological adaptations of honeybees and their surroundings. Every step of the honey-making process, from nectar processing and foraging to comb building and honey ripening, highlights bees' complex behaviors and physiological capacities. Using this scientific knowledge, beekeepers can better manage their hives, produce honey of higher quality, and promote sustainable beekeeping methods that preserve bees' essential role as pollinators and producers of nature's miracle food. Beekeepers ensure a sweet future for bees and humans by protecting bee populations, biodiversity, and global food security via ongoing research, teaching, and innovation.

Pollination and its impact on agriculture

Pollination is a fundamental biological process primarily carried out by bees, insects, birds, and the wind. It has significant effects on agriculture, biodiversity, and food security. In this natural event, pollen grains are transferred from the male (anther) to the female (stigma) section of flowers, allowing flowering plants to fertilize and produce fruit and seeds.

Pollination is essential to preserving biodiversity and the balance of ecosystems. As the principal pollinators, bees visit flowers in quest of nectar and pollen, unintentionally spreading pollen from one bloom to another. Numerous plant species, including many crops and wild plants vital to wildlife habitats, are supported in their reproduction by this process. Pollination increases a plant's resistance to environmental stressors, increases genetic variety within plant populations, and promotes ecosystem stability.

Pollination is essential to the growth of many crops in agriculture, which provide substantial food for people and means of subsistence. "Quality and production of crops": Crop productivity, fruit set, and quality characteristics, including size, shape, and homogeneity, are all directly impacted by Pollination.

"Bee-pollinated" crops include almonds, apples, berries, and cucurbits (cucumbers, melons, and squash), which depend primarily on bees and other insect pollinators for adequate fertilization and bountiful harvests. For instance, bees must cross-pollinate almond trees, a vital commodity for California's agricultural economy, to produce almonds.

"Cost Effects:" insect pollinators contribute billions of dollars each year to worldwide agricultural output each year, demonstrating the significant economic worth of pollination services. Pollinators contribute to farm revenues, rural livelihoods, agricultural resilience to environmental issues, and immediate financial benefits. Farmers and growers collaborate with beekeepers and environmentalists to maximize farm pollination and protect pollinator health. They acknowledge bees' critical role in maintaining productive fields, orchards, and greenhouse crops.

Pollination is crucial, but it faces many obstacles that endanger pollinator populations and the sustainability of agriculture. Pollination efficiency and agricultural output are compromised by "habitat loss," "pesticide exposure," "climate change," and "disease" outbreaks, which can jeopardize bee health and foraging behavior. To promote pollinator populations and maintain productive agricultural landscapes, beekeepers and agricultural stakeholders work together to develop "pollinator-friendly practices" such as "habitat restoration," "pesticide risk mitigation," and "sustainable farming methods."

Effective pollinator conservation and management measures are crucial to protect pollination services and increase agricultural resilience. The main goals of conservation are to maintain a variety of " pollinator habitats," to encourage the growth of "native plant species" that benefit pollinators, and to develop "pollinator-friendly landscapes" in urban and agricultural settings. Beekeepers play a critical role in pollinator management by keeping healthy bee colonies, strategically placing hives to maximize foraging possibilities, and lobbying for laws that put pollinator health and biodiversity conservation first.

Innovation in sustainable agriculture and pollinator management is fueled by ongoing study into pollinators' biology, behavior, and ecosystem interactions. To create "best practices" for improving pollination services and reducing risks, scientists investigate "genetic diversity" in pollinator populations, "pollination efficiency," and "crop-pollinator interactions." Technological developments in "precision agriculture," "pollen analysis," and "remote sensing" allow farmers and beekeepers to maximize crop pollination, monitor pollinator activity, and make well-informed decisions that support pollinator health and agricultural productivity.

Pollination impacts environmental sustainability and global food security in addition to specific farms and geographic areas. To ensure a "secure food supply," "biodiversity conservation," and "ecosystem resilience" in the face of global problems, including population growth, climate variability, and land use change, it is imperative to address pollinator decreases and promote robust pollination networks. To encourage a "coordinated approach" to pollinator protection and sustainable agriculture, cooperation between governments, the agricultural sector, beekeepers, and conservation organizations is essential.

Pollination is essential to global ecosystem health, biodiversity, and agricultural productivity. Stakeholders can protect pollination services, support sustainable agriculture, and guarantee a resilient future for food production and natural ecosystems by acknowledging the importance of pollinator contributions, encouraging pollinator-friendly practices, and advancing scientific understanding and Innovation. We uphold the vital role bees, and other pollinators play in supporting life on Earth by working together to safeguard pollinators and their habitats, promoting harmony between agriculture and nature for future generations.

Understanding bee genetics and breeding

Beekeepers must understand bee genetics and breeding to increase colony health, production, and resilience against diseases and environmental stressors. The European honeybee (Apis mellifera), in particular, is known for its complex genetic features affecting its behavior, disease resistance, ability to produce honey, and general colony dynamics.

The study of bee genetics involves various characteristics passed down through mating between queens, female bees, drones, or male bees. A single queen bee who lays eggs, drones that contribute to genetic variation through mating flights, and worker bees—female bees who carry out different activities within the hive—typically comprise each bee colony. Genetic variety among colonies ensures resilience against environmental changes, pests, and diseases, which is essential for sustaining healthy and productive bee populations.

Beekeepers use selective breeding programs to transmit desired features in populations of honeybees. "docility," "honey production efficiency," "disease resistance," and "temperament" are possible, desirable qualities. Beekeepers strive to improve their colonies' overall performance and capacity to adapt to the local environment by carefully selecting queens from colonies that possess these characteristics. To maintain sustainable beekeeping practices and increase agricultural production, selective breeding also helps to create bee populations that flourish in some geographic regions or rural contexts.

Various breeding techniques transmit desired bee traits. "Instrumental insemination efficiently" allows beekeepers to manage mating and add particular genetic traits to bee populations. Selecting queens and drones based on observed attributes like gentleness, brood viability, or honey output is known as "artificial selection." This strategy benefits breeding projects focusing on disease resistance or behavior modification. In regulated settings, natural mating maintains genetic variety and lowers the possibility of introducing unwanted traits into bee populations.

Developments in this field have led to a better understanding of the biology and behavior of honeybees, which has influenced management strategies and breeding initiatives. "Inheritance patterns" of significant features, "genetic markers" linked to disease resistance, and "population dynamics" within bee colonies are all clarified by genetic research. Researchers investigate the genetic basis of complex features and devise plans for improving bee health, productivity, and resilience in the face of environmental difficulties using molecular tools such as " DNA sequencing" and "genomic analysis."

Beekeepers struggle to manage bee genetics and breeding programs. In small populations, "genetic drift" can happen, resulting in decreased genetic diversity and increased vulnerability to illnesses or environmental stressors. To prevent unforeseen outcomes like inbreeding depression or the loss of adaptive features, it is essential to strike a balance between the maintenance of genetic variety and desirable qualities. Together, beekeepers and "breeding associations," "research institutions," and "extension services" provide access to genetic resources, exchange breeding techniques, and advance sustainable beekeeping best practices.

Bee breeding and genetics are significantly advanced by community involvement. To share information, obtain breeding stock, and work together on genetic research projects, beekeepers participate in " breeding networks," "queen-rearing workshops," and "genetic improvement programs." By increasing public knowledge of the value of gene diversity, ethical breeding methods, and the function of bees in agricultural ecosystems, educational outreach programs encourage a spirit of creativity and responsibility among stakeholders and beekeepers.

To handle new issues like "climate change," "pesticide exposure," and "pathogen resistance," bee genetics and breeding have a bright future. "Resilience traits" that improve bee health and survival in dynamic environmental situations are studied by researchers. Combining "genomic selection" with "precision breeding" methods speeds up attempts to enhance genetics, allowing beekeepers to control bee populations adaptively and support sustainable agriculture. Beekeepers support robust bee populations, biodiversity conservation, and global food security through scientific breakthroughs and cooperative partnerships.

To sum up, maintaining healthy bee populations, increasing agricultural production, and reducing dangers to pollinator health depend on understanding bee genetics and breeding. Beekeepers are essential to preserving bees' critical roles in ecosystems and food systems worldwide because they prioritize genetic variety, selective breeding, and scientific innovation. Using continuous investigation, instruction, and cooperative endeavors, interested parties preserve the heritage of conscientious beekeeping methods, guaranteeing a robust future for both hives and human communities.

Advances in bee research and technology

Our knowledge of bee behavior, health, and interactions with human activity has been entirely transformed by bee science and technology advances. Bees play an essential part in ecosystems. These developments cover various fields, including ecology, conservation, genetics, physiology, and ecology. They provide important new information for pollinator conservation initiatives and sustainable beekeeping methods.

Bee genetics is a critical area of progress. Presently, scientists are employing sophisticated "genomic techniques," such as "DNA sequencing," to decipher the genetic underpinnings of significant features like behavior, illness resistance, and honey output. By enhancing desired features in bee populations and strengthening their resistance to illnesses and environmental stressors, selective breeding methods made possible by an understanding of bee genetics can be implemented. Techniques such as "genomic selection" and "marker-assisted selection" are being employed more often to speed up the breeding of bees with features useful for both conservation and commercial objectives.

With growing worries about "pesticides," "parasites" such as the "Varroa mite," and "pathogens" like "viruses" and "Nosema," advancements in "bee health" have been crucial. Researchers have created "diagnostic tools" to identify these risks early that enable beekeepers to apply focused "treatments" and "management strategies." Advances in "biotechnology" and "bioinformatics" contribute to developing "vaccines" and "treatments" customized to address certain bee diseases, protecting bee populations, and maintaining their pollinator role.

Technology has revolutionized "management" and "hive monitoring." Temperature, humidity, and hive weight are among the elements of hive conditions that are remotely monitored using "remote sensing," "sensor networks," and "Internet of Things" devices. These tools give beekeepers access to real-time data that they can use to forecast swarming tendencies, optimize feeding schedules, and harvest honey at the ideal moments. "Drone technology" is also being investigated for labor-intensive beekeeping chores like queen monitoring and hive inspections.

Pollination ecology research has shown how vital bees are to "crop pollination" and ecosystem health. Improvements in our knowledge of bee preferences for flowers, foraging behavior, and pollination efficiency enable us to maximize "crop yields" and "biodiversity conservation." Policymakers and stakeholders can be guided toward sustainable practices using "modeling techniques" and "ecological studies" to forecast how land use, climate, and pesticide use affect bee populations and pollination services.

Developments in bee research are propelling "pollinator conservation "through "community engagement" and "policy changes." "Reduction of pesticide use," "support for bee-friendly agriculture," and "habitat preservation" are promoted by research findings on the "ecological services" rendered by bees. Campaigns for public awareness and educational initiatives help people understand the value of bees and provide them with the tools they need to protect pollinators in their neighborhoods.

Technological advancements in bee research and applications require international collaboration. Through "global networks" and "research partnerships," scientists, beekeepers, and policymakers work together to exchange best practices, resources, and expertise. The "global challenges" that this partnership addresses, like "bee declines," "invasive species," and "climate change impacts" on bee habitats and pollination services, will be addressed more quickly.

There is hope for more innovation and discovery in bee science and Technology. Future "genetic research" might identify novel "traits" that improve "the adaptability" and "resilience" of bees in dynamic environments. More "efficient" and "sustainable beekeeping practices" will probably result from "technological advancements," which will help "small-scale beekeepers" throughout various regions and commercial beekeepers. "Policy integration" of research findings can lead to "regulatory actions" that support "biodiversity conservation" worldwide and bee-friendly practices.

In summary, bee science and technology developments are revolutionizing our capacity to comprehend, control, and preserve bee populations. Through pollination services, these innovations guarantee food security, promote sustainable farming practices, and aid in the global conservation of biodiversity. We can ensure that bees' essential contributions to ecosystems and human well-being are preserved for future generations by utilizing scientific innovation, encouraging cooperation, and involving communities and governments.

CHAPTER IX

The Art of Beekeeping

Building a bond with your bees

The art of beekeeping transcends mere agricultural practice; it is a relationship built on understanding, respect, and stewardship between beekeepers and their colonies. At its core, beekeeping involves not just the management of hives for honey production but also nurturing and fostering a harmonious bond with the bees themselves.

Understanding their behavior and communication is central to bonding with bees. Bees operate as a highly organized society with distinct roles: the queen, workers, and drones. Each plays a crucial part in the hive's functionality, from the queen's egg-laying duties to the workers' meticulous tasks of foraging, nursing the brood, and maintaining the hive's temperature and cleanliness. Drones contribute to genetic diversity through mating flights. Beekeepers who grasp these roles and behaviors can anticipate hive needs and respond accordingly, fostering a sense of security and stability within the colony.

The art of beekeeping embodies a profound respect for the natural world and its rhythms. Beekeepers observe seasonal cycles, floral blooms, and environmental conditions influencing bee activity and health. They adapt management practices to support bees through periods of nectar dearth, harsh weather, or potential pests and diseases. By aligning with nature's rhythms, beekeepers promote hive well-being and cultivate a deeper connection to the ecosystems that sustain their bees.

Building a bond with bees requires patience and keen observation. Beekeepers learn to interpret the subtle cues and behaviors exhibited by their colonies. They observe the intensity and direction of bee traffic at hive entrances, the hum of activity within the hive, and the dance-like movements of foraging bees communicating the location of rich nectar sources. This attentive observation enables beekeepers to intervene thoughtfully while respecting the hive's natural processes and rhythms.

Beekeeping is often associated with the image of beekeepers clad in protective gear, gently tending to their hives amidst a backdrop of buzzing activity. A gentle and calm approach is vital to building trust and minimizing stress among bees. Beekeepers move slowly and deliberately during hive inspections, using smoke to calm bees and minimize defensive behaviors. This approach ensures the safety of both bees and fosters a relaxed atmosphere conducive to productive hive management.

Beekeeping involves continuous learning and skill development. Successful beekeepers stay informed about advancements in bee health, genetics, and management practices. They attend workshops, participate in beekeeping associations, and engage with experienced mentors to deepen their knowledge and craft. Through education, beekeepers gain practical insights into hive management techniques, pest and disease control strategies, and sustainable beekeeping practices that promote colony longevity and productivity.

Ethical considerations are integral to beekeeping. Responsible beekeepers prioritize the well-being of their bees above all else. They adhere to organic and sustainable beekeeping practices, avoiding harmful chemicals and antibiotics that may compromise bee health or contaminate honey. Ethical beekeeping also involves providing bees with adequate forage, clean water sources, and optimal hive conditions that support their natural behaviors and life cycles.

The art of beekeeping offers rewards beyond harvesting honey and other hive products. For many beekeepers, the connection forged with their bees is deeply fulfilling. It fosters a sense of stewardship and responsibility towards the environment and pollinators crucial for agriculture and biodiversity. Beekeepers often profoundly respect bees' resilience, intelligence, and contributions to ecosystems, inspiring a commitment to sustainable practices that benefit both bees and their broader communities.

Building a bond with bees extends to fostering community among fellow beekeepers and advocating for pollinator conservation. Beekeeping associations and local groups provide platforms for sharing knowledge, exchanging resources, and supporting novice beekeepers. Through advocacy efforts, beekeepers raise awareness about the importance of bees as pollinators and the threats they face, mobilizing communities to protect habitats, reduce pesticide use, and support bee-friendly practices in agriculture and urban landscapes.

In conclusion, the art of beekeeping is a multifaceted practice that blends science, stewardship, and respect for nature. Beekeepers nurture strong bonds with their colonies by cultivating a deep understanding of bee behavior, practicing patience and observation, and embracing ethical practices. This bond enhances hive health and productivity and fosters a profound connection to the natural world and a commitment to safeguarding pollinators for future generations. Through dedication, education, and community engagement, beekeepers are vital in promoting sustainable agriculture, biodiversity conservation, and preserving bees' essential role in global ecosystems.

Observing and interpreting bee behavior

Observing and interpreting bee behavior is a fundamental skill for beekeepers, vital for understanding colony health, anticipating hive needs, and fostering a harmonious relationship with bees. As social insects, Bees exhibit intricate behaviors that communicate their needs, roles within the hive, and responses to environmental stimuli.

One of the most visible and crucial behaviors to observe is foraging. Worker bees embark on foraging flights to collect nectar, pollen, water, and propolis essential for hive nutrition and maintenance. Beekeepers can observe the intensity and direction of bee traffic at hive entrances, indicating floral resources' abundance and proximity. By monitoring foraging behavior, beekeepers gauge the health of surrounding floral landscapes and assess the nutritional status of their colonies.

Honeybees are renowned for their waggle dance, a complex form of communication that conveys the location, quality, and distance of food sources to nestmates. Observing dance language allows beekeepers to deduce the types of flowers bees visit, their travel distances, and the quality of nectar or pollen being brought back to the hive. This information informs beekeepers about seasonal changes in forage availability and helps optimize hive placement to maximize foraging efficiency.

Bee colonies may exhibit behaviors signaling their readiness to swarm, a natural reproductive process where a portion of the colony leaves with a new queen to establish a new hive. Signs such as increased drone production, queen cell construction, and queen slimming (where workers feed the queen less) indicate swarm preparations. By recognizing these behaviors early, beekeepers can implement management techniques to prevent colony loss and capture swarms for expansion or beekeeping education.

Bees exhibit defensive behavior when they perceive threats to the hive, such as predators, disturbances, or intrusions by beekeepers during hive inspections. Defensive behaviors include buzzing loudly, flying in a defensive formation around the hive entrance, and stinging to protect the colony. Beekeepers learn to approach hives calmly and use smoke to pacify bees, minimizing defensive responses and ensuring both beekeeper and bee safety during hive manipulations.

Workers in the hive perform essential tasks such as nursing the brood (developing larvae and pupae) and maintaining nest hygiene. Observing these behaviors offers insights into colony vitality and queen health. Worker bees meticulously clean cells, remove debris, and regulate hive temperature and humidity to create optimal conditions for brood development. Changes in brood care behaviors, such as erratic nursing or increased cell capping, may signal queen issues or disease challenges requiring beekeeper intervention.

Bee behavior varies seasonally in response to changes in temperature, daylight length, and floral availability. During colder months, bees cluster tightly to conserve warmth and maintain hive temperature. As spring approaches, bees increase foraging activities to capitalize on blooming flowers and replenish hive stores. Observing seasonal variations in bee behavior allows beekeepers to anticipate colony needs, provide supplementary feeding when necessary, and adjust management practices to support hive survival and productivity throughout the year.

Successful beekeeping hinges on continuous learning and adaptation to evolving behaviors and environmental conditions. Beekeepers engage in ongoing education through workshops, mentorships, and beekeeping associations to deepen their understanding of bee behavior and refine their observational skills. By staying informed about advancements in bee research and management practices, beekeepers can implement evidence-based strategies that promote colony health, resilience, and sustainable beekeeping practices.

Observing bee behavior also underscores the importance of ethical beekeeping practices and environmental stewardship. Beekeepers prioritize providing bees with diverse forage, minimizing hive disturbances, and avoiding the use of harmful chemicals that can impact bee health and the environment. By fostering a respectful and empathetic approach to beekeeping, beekeepers contribute to the conservation of pollinator habitats and promote the essential role of bees in sustaining biodiversity and agricultural productivity.

In conclusion, observing and interpreting bee behavior is both an art and a science essential to successful beekeeping. By honing their observational skills and understanding the nuances of bee communication, beekeepers can enhance colony management, support pollinator health, and contribute to sustainable agricultural practices. Through careful observation, continuous learning, and ethical stewardship, beekeepers play a vital role in safeguarding bees' contributions to ecosystems and food systems worldwide.

Ethical beekeeping practices

Ethical beekeeping practices encompass principles and guidelines that prioritize the well-being of bees, promote sustainable hive management, and uphold environmental stewardship. These practices are rooted in respect for the natural behavior and needs of honeybees (Apis mellifera) and other pollinators crucial to global ecosystems and agricultural productivity.

Ethical beekeepers prioritize providing bees access to diverse and abundant forage sources yearly. This includes planting bee-friendly flowers, trees, and shrubs that offer nectar and pollen essential for bee nutrition. By supporting diverse floral landscapes, beekeepers enhance bee health, promote colony resilience, and contribute to the conservation of pollinator habitats. Ensuring adequate forage helps mitigate nutritional stress and supports bees during times of nectar dearth.

Ethical beekeepers favor natural and minimalist hive management approaches that align with bees' natural behavior and life cycles. This includes providing bees with a natural comb foundation rather than synthetic materials, allowing bees to build comb according to their needs. Ethical beekeeping practices prioritize minimal intervention during hive inspections, reducing disturbances to bee colonies and allowing bees to regulate hive conditions independently.

Promoting bee health is a cornerstone of ethical beekeeping. Beekeepers regularly monitor hive health, observing bee behavior, brood patterns, and hive conditions to detect early signs of disease or parasitic infestations. Ethical beekeepers employ integrated pest management (IPM) strategies, emphasizing prevention, monitoring, and non-chemical control methods to manage pests like the Varroa mite and diseases like Nosema. Ethical beekeepers prioritize organic and bee-friendly options when treatments are necessary to minimize harm to bees and hive products.

Ethical beekeepers engage in selective breeding programs to propagate bees with desirable traits such as gentleness, disease resistance, and honey production efficiency. They prioritize genetic diversity to enhance colony resilience against environmental stressors and diseases. Ethical beekeeping practices include raising queens from colonies exhibiting desirable traits and avoiding inbreeding to maintain genetic vitality within bee populations. By selecting healthy and productive queens, beekeepers contribute to the long-term sustainability of beekeeping and pollination services.

Ethical beekeeping practices prioritize environmental responsibility and sustainability. Beekeepers avoid placing hives in areas with high pesticide exposure and advocate for reduced pesticide use in agricultural practices. They collaborate with farmers, land managers, and policymakers to promote pollinator-friendly farming practices and habitat conservation. Ethical beekeepers participate in local conservation initiatives and support ecological restoration efforts that benefit bees and pollinators.

Ethical beekeepers prioritize education and community engagement to raise awareness about the importance of bees and pollinators. They share best practices in beekeeping, pollinator conservation, and sustainable agriculture through workshops, public demonstrations, and mentorship programs. By fostering community support and advocacy, ethical beekeepers empower citizens to take positive action to protect pollinators and biodiversity.

Ethical beekeepers prioritize transparency in beekeeping practices and bee product labeling. They ensure honey, beeswax, and other hive products are ethically sourced and sustainably harvested. Beekeepers adhere to food safety and quality standards to deliver pure, unadulterated products to consumers. Ethical beekeepers educate consumers about the value of locally produced and responsibly sourced bee products, fostering a connection between consumers and the environmental benefits of supporting pollinators.

In conclusion, ethical beekeeping practices are grounded in compassion, responsibility, and sustainability. Ethical beekeepers play a vital role in safeguarding bee populations, promoting pollinator diversity, and ensuring the long-term viability of beekeeping and agricultural ecosystems by prioritizing bee health, environmental stewardship, and community engagement. Through dedication to moral principles and continuous learning, beekeepers contribute to a harmonious relationship between humans and bees, benefiting global biodiversity and food security.

Creating a bee-friendly garden

Creating a bee-friendly garden is a delightful endeavor and a crucial step in supporting pollinator health and biodiversity. As primary pollinators, Bees play a fundamental role in ecosystem function and food production, making their conservation and well-being paramount. Designing a garden that caters to bees' needs ensures a thriving habitat where they can forage for pollen and nectar, nest, and contribute to local agricultural productivity.

The foundation of a bee-friendly garden lies in selecting plants that provide abundant and diverse sources of pollen and nectar throughout the growing season. Opt for native plants and wildflowers adapted to your region's climate and soil conditions. Consider planting various flowering species with different bloom times to ensure a continuous food supply for bees from early spring to late fall. Flowers such as lavender, sunflowers, cosmos, bee balm, and coneflowers are beautiful to bees due to their vibrant colors and rich nectar resources.

Create a diverse habitat in your garden to accommodate the needs of different bee species. Incorporate sheltered areas with undisturbed soil** or dead wood for solitary bees that nest in burrows or cavities. Provide water sources such as shallow dishes filled with clean water and pebbles to prevent drowning. Include rock piles or native grasses that offer nesting sites and protection from predators.

To maintain a bee-friendly environment, minimize or eliminate the use of pesticides, herbicides, and synthetic fertilizers that can harm bees and other pollinators. Instead, opt for organic gardening methods and natural pest control solutions to manage garden pests while preserving bee health. Choose resistant plant varieties and companion planting techniques to reduce pest problems naturally without compromising pollinator safety.

Practice seasonal garden maintenance to ensure your bee-friendly garden remains healthy and attractive year-round. Regularly deadhead spent flowers to encourage continuous blooming and maintain plant vitality. Mulch garden beds with organic materials to conserve moisture, suppress weeds, and improve soil health. Monitor plants for signs of disease or nutrient deficiencies and address issues promptly to support robust plant growth and sustained bee foraging opportunities.

Use your bee-friendly garden as an educational tool to raise awareness about the importance of bees and pollinators in the ecosystem. Host educational workshops or garden tours for neighbors, schools, or community groups to share best practices for creating bee-friendly habitats. Provide informational signage, brochures, or online resources highlighting the benefits of pollinator-friendly gardening and inspiring others to adopt similar practices in their yards.

Engage with local beekeeping associations, environmental organizations, and government agencies to collaborate on pollinator conservation initiatives. Participate in citizen science projects or pollinator monitoring programs to contribute valuable data on bee populations and habitat preferences. Advocate for pollinator-friendly policies and land-use practices that protect and enhance bee habitats across urban, suburban, and rural landscapes.

In conclusion, creating a bee-friendly garden is a meaningful way to support pollinator health, biodiversity, and sustainable gardening practices. By choosing bee-attractive plants, providing diverse habitat features, avoiding harmful chemicals, practicing seasonal care, and engaging in education and advocacy efforts, gardeners can play a vital role in promoting the well-being of bees and other essential pollinators. Through collective action and community stewardship, bee-friendly gardens contribute to pollinator species conservation and preserving our natural ecosystems for future generations.

CHAPTER X

Growing Your Beekeeping Skills

Expanding your apiary

Increasing the size of your apiary will help you become a better beekeeper by combining experience, education, and cautious management techniques to maintain the well-being and production of your bee colonies. Many beekeepers are anxious to grow their operations to boost honey output, aid in local pollination efforts, or learn more about the fascinating world of beekeeping as they become more comfortable and proficient in managing their first colonies.

It is essential to evaluate your resources and level of preparation before adding to your apiary. Determine the knowledge, expertise, and free time you must devote to beekeeping. Think about your material and monetary resources, such as tools, safety supplies, and room for more hives. Determine whether the forage in your area can sustain additional bees without endangering the ecology or the health of the bees.

Careful planning is essential for a successful apiary expansion. Create a growth strategy that fits your capabilities and objectives. Decide how many more hives you want to oversee and when you want to expand. Arrange hive placement to reduce rivalry between colonies and maximize accessibility to forage. Ensure you have enough equipment and hive materials ready to accommodate additional colonies.

There are various ways to acquire bees for expansion. Consider buying package bees or nucleus colonies (nucs) from reliable vendors. Nucs are fully formed colonies prepared to grow in a new hive with a queen, brood, and worker bees. Package bees are made up of a queen who must build her comb and brood and a loose cluster of bees. Alternatively, you could split established, robust colonies to form new hives; ensure each split has enough supplies and a mated queen for survival.

Give hive health and management practices top priority as your apiary grows. To monitor colony strength, brood patterns, and disease or pest infestation indications, conduct hive inspections regularly. Use integrated pest management (IPM) techniques to handle illnesses like American foulbrood and pests like the Varroa mite. Ensure there are enough food stores and clean water sources for each hive.

The expansion of apiaries offers chances for ongoing education and skill improvement. Keep up with "new research," best beekeeping practices, and hive management innovations. Join your local **beekeeping associations, attend workshops, and look for beekeeper mentorship to hone your skills and expand your knowledge. Every hive expansion brings fresh difficulties and educational opportunities to further your development as an experienced beekeeper.

Keep well-documented records of every hive's development, including productivity data, queen lineage, hive inspections, and treatments. By keeping track of this data, you can make more educated decisions about genetic selection, hive management, and resource allocation. Maintaining records is also helpful for diagnosis and research during disease outbreaks or problems.

Think about growing your beekeeping skills and services as you grow your apiary. Look into ways to sell honey sales at farm markets or local shops or provide pollination services to nearby farmers. Spread your love of beekeeping by informing people, giving hive tours, or organizing beekeeping workshops for newbies.

Growing apiaries can promote community involvement and awareness of the importance of pollinators and bees in the community. Encourage your neighbors and local schools to see your apiary and learn about the advantages of pollinator conservation and bee-friendly gardening. Work with neighborhood organizations to advance pollinator-friendly policy and sustainable beekeeping practices.

Growing your apiary is an enjoyable experience that calls for cautious preparation, ongoing education, and a commitment to hive health and sustainability. You can support pollinator conservation and local ecosystem health by responsibly acquiring bees, practicing excellent management techniques, and cultivating a community of beekeepers. Developing your beekeeping skills will help you preserve agricultural biodiversity and bee populations for future generations while strengthening your bond with nature.

Joining beekeeping communities and networks

Beekeepers can take advantage of priceless chances for learning, cooperation, and support from one another in the ever-changing world of apiculture by becoming members of beekeeping communities and networks. Even though beekeeping is incredibly fulfilling, it can also be complicated and complex, so community involvement is crucial for both beginning and seasoned beekeepers.

The chance to learn and share is one of the most significant advantages of becoming involved in beekeeping groups. Beekeepers share knowledge, strategies, and helpful tips on managing hives, controlling pests, producing honey, and other topics via local beekeeping associations, clubs, or online forums. Years of practical experience have resulted in collective wisdom that aids beekeepers in overcoming obstacles and implementing creative solutions that improve hive health and output.

Beekeeping associations frequently host workshops, seminars, and training sessions" led by knowledgeable entomologists, farmers, and beekeepers. These instructional programs cover many topics, from fundamental beekeeping methods to sophisticated hive management techniques and the most recent findings in bee health and sustainability. By participating in these programs, beekeepers can acquire the information and abilities to overcome challenges, adjust to shifting environmental conditions, and encourage ethical beekeeping practices.

Participating in beekeeping groups opens up mentoring chances with other beekeepers, scientists, and business experts. Beekeepers can get mentorship from experienced professionals who can advise, troubleshoot issues, and offer encouragement along the beekeeping journey by building connections within the community. Beyond providing technical guidance, mentoring relationships frequently promote camaraderie and community among beekeepers.

Beekeeping communities are hubs for the lending and exchanging of tools, parts for hives, and genetic material, such as queens or nucleus colonies (nucs). By lowering waste and encouraging effective resource allocation within the community, this cooperative method encourages sustainable practices and makes it more affordable for beekeepers to obtain the necessary resources.

At the local, regional, and federal levels, beekeeping communities are essential in promoting pollinator protection, sustainable agriculture, and bee-friendly policies. Beekeepers can lobby for policies that protect bee habitats and forage resources, influence lawmakers, and educate the public about bees' vital role in ecosystem health and food security.

Beyond merely possessing technical know-how, involvement in beekeeping communities promotes community engagement and public education regarding bees and pollinators. Beekeepers frequently participate in community events, school presentations, and farmers' markets to spread awareness about pollinator conservation, sustainable gardening, and the advantages of locally produced honey and hive products. These educational initiatives encourage others to take proactive measures to support pollinator health and advance a more profound knowledge of ecological interdependence.

Beekeeping groups honor variety in apiculture-related beekeeping methods, hive types, and cultural traditions. Accepting different viewpoints and methods adds to the body of knowledge, fosters creativity, and stimulates inclusive beekeeping activities. Beekeeping communities enable beekeepers from many backgrounds to provide their distinct perspectives and experiences toward sustainable beekeeping by creating a warm and inviting atmosphere.

Finally, the beekeeping journey is enhanced by knowledge sharing, educational opportunities, mentorship, resource sharing, advocacy, and community engagement when beekeepers join beekeeping communities and networks. In addition to developing their abilities and resiliency, beekeepers actively participate in these networks to support bee habitat preservation, sustainable agriculture, and pollinator conservation. Beekeeping communities create a culture of innovation, collective action, and stewardship via cooperation and mutual support, ensuring the health of our ecosystems and the vitality of bee populations for future generations.

Continuing education and resources

Beekeepers need access to resources and ongoing education at every level of their career to stay knowledgeable, skilled, and proactive in managing their apiaries. As a dynamic field affected by environmental changes, scientific discoveries, and emerging best practices, beekeeping requires constant learning to keep colonies healthy and maximize production.

Participating in continuing education programs catering to their needs and interests enormously benefits beekeepers. Local beekeeping associations, universities, and industry professionals host workshops, seminars, and conferences that provide comprehensive insights into advanced hive management practices, pest and disease control measures, and the most recent findings in bee health and genetics. Through these instructional forums, beekeepers can improve their general proficiency in beekeeping operations and gain practical skills and knowledge to solve issues effectively.

Beekeepers may now access a plethora of knowledge from anywhere in the world thanks to the revolutionary changes in beekeeping education brought about by the accessibility of online resources. Many topics are covered in online courses, webinars, and instructional videos, ranging from the fundamentals for beginners to more specialist areas like raising queens, building hives, and organic beekeeping techniques. Through forums and social media groups, digital platforms also provide peer-to-peer learning, enabling beekeepers to instantly share success stories, troubleshoot issues, and discuss ideas.

A wealth of material about beekeeping is available to beekeepers, ranging from in-depth manuals and guides to academic publications and magazines. Books written by well-known authorities on beekeeping offer in-depth analyses of bee biology, hive management concepts, and helpful guidance on seasonal hive tasks. Beekeepers can stay updated on legislative developments, industry trends, and scientific advancements affecting their practices and policies by subscribing to beekeeping magazines.

Local resources are essential to beekeepers' success and continuous education. One can obtain significant opportunities for networking, practical training, and mentorship from seasoned beekeepers through local beekeeping clubs, associations, and mentorship programs. Beekeepers benefit from these grassroots initiatives by developing a sense of camaraderie among themselves, exchanging knowledge, working together to solve problems, and cultivating sustainable beekeeping techniques adapted to local climates and environmental conditions.

Keeping up with the latest developments in beekeeping science and technology is crucial for making well-informed decisions and ensuring that apiary management is continuously improved. Research on the behavior, genetics, and health of bees sheds light on new issues such as colony collapse disorder and the effects of pesticides on pollinators. Beekeeping equipment, sustainable beekeeping methods, and hive design innovations have brought new tools and techniques to maximize hive output while reducing environmental effects.

Successful beekeeping is characterized by its capacity to adjust to shifting environmental conditions and unforeseen obstacles. Beekeepers who get ongoing instruction are better equipped to foresee seasonal variations, prevent hive illnesses, and adapt to changing climate conditions and habitat loss. Beekeepers contribute to bee populations' long-term health and sustainability by utilizing local resources and incorporating knowledge from various educational sources to create resilience in their apiaries.

To sum up, access to resources and continual education are essential to the resilience and success of beekeepers worldwide. Beekeepers acquire the information, abilities, and assistance required to successfully negotiate the intricacies of contemporary beekeeping methods through professional development opportunities, online learning environments, literature, local community involvement, and breakthroughs in research and innovation. Beekeepers support sustainable agriculture, bees' crucial role in global ecosystems, and the preservation of pollinator health by encouraging a culture of lifelong learning and proactive management.

Future trends in beekeeping

The future of beekeeping is bright, with opportunities and difficulties brought forth by changing agricultural practices, environmental changes, and technological improvements. Beekeeping is a critical practice shaped by numerous major trends as academics and beekeepers continually develop new ideas and ways to adapt.

Technology integration has the potential to completely transform beekeeping operations by providing new instruments for data collecting, disease diagnosis, and hive management. Sensor-equipped innovative hive technology allows real-time monitoring of hive parameters like temperature, humidity, and weight. Beekeepers can make well-informed decisions at a distance with the help of these data points, resulting in optimal hive health and productivity. Drones with imaging and mapping capabilities help track pollen distribution, hive inspections, and bee behavior. They also shed light on colony dynamics and the effects they have on the environment.

Bee breeding systems are being improved by genetic research advances to create robust and productive bee stocks. Selective breeding for features including illness resistance, productivity in producing honey, and behavioral traits help create robust populations of honeybees that can survive in various environmental situations. Understanding bee genetics also makes identifying genetic markers linked to desired features easier, opening the door to more focused breeding techniques and the preservation of genetic diversity.

Pollinator health, including bee health, is still a global problem because of habitat loss, chemical exposure, climate change, and disease stresses. Beekeeping's future trends will focus on sustainable methods that prioritize pollinator health and habitat preservation. In an effort to reduce environmental stressors and encourage pollinator-friendly landscapes, integrated pest management techniques, organic beekeeping methods, and habitat restoration projects are becoming more and more popular. To protect pollinator populations and promote biodiversity, beekeepers, farmers, academics, and legislators must work together to implement evidence-based solutions.

Growing awareness of pollinator conservation, local food production, and urban agricultural efforts are driving the emergence of urban beekeeping as a trend. To take advantage of the variety of foraging options and lessen the adverse effects of urbanization on bee habitats, beekeepers are setting up hives in urban settings like roofs, community gardens, and urban green spaces. In addition to helping the local economy by producing honey and providing pollination services, urban beekeeping encourages community involvement and raises awareness of the importance of bees in urban ecosystems.

Beekeeping industry trends are being impacted by growing consumer knowledge of the value of pollinators and sustainable food systems. Demand for handmade hive goods, locally produced honey, and bee-friendly farming methods is rising. In response to customer demands, beekeepers are embracing transparent beekeeping procedures, supporting the health and welfare of honeybees, and pushing for laws that favor pollinators. This consumer-driven movement upholds the importance of bees in sustainable agriculture and the health of ecosystems while assisting in the financial sustainability of beekeeping businesses.

The future of beekeeping is greatly influenced by education and advocacy initiatives, which enable beekeepers, legislators, and the general public to take proactive measures for pollinator conservation. Beekeeping groups, academic institutions, and charitable organizations are essential to promote best practices, research, and push for laws that safeguard pollinators and their habitats. Diverse stakeholders are involved in cooperative efforts to solve issues confronting bee populations and promote a culture of stewardship toward pollinator health through outreach programs, beekeeping workshops, and citizen science activities.

To sum up, beekeeping has a bright future filled with opportunities for creativity, sustainability, and adaptability to changing environmental conditions. Technological advancements, genetic research, sustainable practices, urban beekeeping efforts, consumer awareness, and education drive revolutionary advances in beekeeping methods and legislation worldwide. Beekeepers can support sustainable agriculture, the preservation of pollinator variety, and the critical role bees play in ecosystem resilience and global food security by adopting these trends and encouraging collaboration across sectors.

CONCLUSION

" Buzzing Beginnings: A Beginner's Guide to Beekeeping: Learn the Art and Science of Keeping Bees in Your Backyard " is an invaluable resource for aspiring beekeepers embarking on their journey into the fascinating world of apiculture. Throughout this book, readers have been equipped with essential knowledge and practical insights into the art and science of beekeeping, from understanding the anatomy and life cycle of bees to learning about hive management, honey harvesting, and the critical role of bees in ecosystems.

By delving into hive setup, bee health, and legal considerations, this guide has provided a comprehensive foundation for beekeepers to establish and maintain thriving colonies responsibly. It emphasizes the importance of sustainable practices, respect for bees' natural behaviors, and engagement with local communities and environments.

Moreover, "Buzzing Beginnings" encourages a deeper appreciation for the intricate relationship between humans and bees, highlighting the benefits of beekeeping beyond honey production, including pollination support for agriculture and conservation efforts for bee populations worldwide.

As readers conclude their journey through this guide, they are empowered with practical skills and knowledge and a sense of stewardship toward bees and their habitats. Whether starting as backyard enthusiasts or aspiring to expand into larger-scale beekeeping operations, "Buzzing Beginnings" equips beekeepers with the tools and understanding needed to embark confidently on their beekeeping adventure, fostering a sustainable and rewarding experience for both bees and beekeepers alike.

Thank you for buying and reading/ listening to our book. If you found this book useful/ helpful please take a few minutes and leave a review on the platform where you purchased our book. Your feedback matters greatly to us.